# SOLVING ENGINEERING PROBLEMS IN DYNAMICS

By

## MICHAEL B. SPEKTOR

Industrial Press

A full catalog record for this book will be available from the Library of Congress

ISBN 978-0-8311-3494-5

**Industrial Press, Inc.**
32 Haviland Street
South Norwalk, Connecticut 06854

Sponsoring Editor: Jim Dodd
Interior Text and Cover Design: Janet Romano
Developmental Editor: Robert Weinstein

**To My Family**

# TABLE OF CONTENTS

*Introduction*                                                                    ix

## Chapter 1   Differential Equations of Motion                                    1

**1.1**   The Left Side of Differential Equations of Motion
          (Sum of Resisting Loading Factors Equals Zero)                           2
**1.2**   The Left and Right Sides of Differential Equations of Motion
          (Sum of Resisting Loading Factors Equals Sum of Active
          Loading Factors)                                                         9

## Chapter 2   Analysis of Forces                                                  17

**2.1**   Analysis of Resisting Forces                                            17
          *2.1.1  Forces of Inertia*                                              18
          *2.1.2  Damping Forces*                                                 19
          *2.1.3  Stiffness Forces*                                               23
          *2.1.4  Constant Resisting Forces*                                      28
          *2.1.5  Friction Forces*                                                28
**2.2**   Analysis of Active Forces                                               29
          *2.2.1  Constant Active Forces*                                         29
          *2.2.2  Sinusoidal Active Forces*                                       29
          *2.2.3  Active Forces Depending on Time*                                29
          *2.2.4  Active Forces Depending on Velocity*                            30
          *2.2.5  Active Forces Depending on Displacement*                        30

## Chapter 3    Solving Differential Equations of Motion Using Laplace Transforms                                   33

**3.1**    Laplace Transform Pairs for Differential Equations of Motion    34

**3.2**    Decomposition of Proper Rational Fractions    40

**3.3**    Examples of Decomposition of Fractions    41

**3.4**    Examples of Solving Differential Equations of Motion    46

    *3.4.1 Motion by Inertia with no Resistance*    46

    *3.4.2 Motion by Inertia with Resistance of Friction*    48

    *3.4.3 Motion by Inertia with Damping Resistance*    49

    *3.4.4 Free Vibrations*    51

    *3.4.5 Motion Caused by Impact*    55

    *3.4.6 Motion of a Damped System Subjected to a Time Depending Force*    59

    *3.4.7 Forced Motion with Damping and Stiffness*    63

    *3.4.8 Forced Vibrations*    65


## Chapter 4    Analysis of Typical Mechanical Engineering Systems                                   69

**4.1**    Lifting a Load    70

    *4.1.1 Acceleration*    71

    *4.1.2 Braking*    76

**4.2**    Water Vessel Dynamics    80

**4.3**    Dynamics of an Automobile    83

    *4.3.1 Acceleration*    84

    *4.3.2 Braking*    93

**4.4.**    Acceleration of a Projectile in the Barrel    95

**4.5**    Reciprocation Cycle of a Spring-loaded Sliding Link    100

    *4.5.1 Forward Stroke Due to a Constant Force*    101

*4.5.2  Forward Stroke Due to Initial Velocity*                        104

*4.5.3  Backward Stroke*                                               108

**4.6**   Pneumatically Operated Soil Penetrating Machine             110


## Chapter 5   Piece-Wise Linear Approximation                        **115**

**5.1**   Penetrating into an Elasto-Plastic Medium                    117

*5.1.1  First Interval*                                               119

*5.1.2  Second Interval*                                              122

*5.1.3  Third Interval*                                               124

*5.1.4  Fourth Interval*                                              126

**5.2**   Non-linear Damping Resistance                               129

*5.2.1  First Interval*                                               131

*5.2.2  Second Interval*                                              133


## Chapter 6   Dynamics of Two-Degree-of-Freedom Systems   **137**

**6.1**   Differential Equations of Motion: A Two-Degree-of-
Freedom System                                                        140

*6.1.1  A System with a Hydraulic Link (Dashpot)*                     140

*6.1.2  A System with an Elastic Link (Spring)*                       142

*6.1.3  A System with a Combination of a Hydraulic Link
(Dashpot) and an Elastic Link (Spring)*                               144

**6.2**   Solutions of Differential Equations of Motion for
Two-Degree-of-Freedom Systems                                         145

*6.2.1  Solutions for a System with a Hydraulic Link*                 145

*6.2.2  Solutions for a System with an Elastic Link*                  148

*6.2.3  Solutions for a System with a Combination
of a Hydraulic and an Elastic Link*                                   152

6.2.4 *A System with a Hydraulic Link where the First
Mass Is Subjected to a Constant External Force*        158

6.2.5 *A Vibratory System Subjected to an External
Sinusoidal Force*        163

**Index**        169

# INTRODUCTION

Purposeful control and improvement of how existing mechanical systems perform is an important real-life problem, as is the development of new systems. We can obtain solutions to these problems by investigating the working processes of machines and their units and elements. These investigations should be based on fundamentals of dynamics combined with a variety of related sciences. The working processes that characterize system performance can be described by mathematical expressions that actually represent equations of motion of these systems. Analyzing these equations of motion reveals the relationship between the parameters of the system and their influence on performance and other system characteristics or elements.

This book contains comprehensive methods for analyzing the motion of engineering systems and their components. The analysis covers three basic phases: 1) composing the differential equation of motion, 2) solving the differential equation of motion, and 3) analyzing the solution. Engineering education provides the fundamental skills for completing these three phases. However, many engineers would benefit from additional training in using these fundamentals to solve real-life engineering problems. This book provides this training by describing in a step-by-step order the methods related to each of these three phases.

When assembling a differential equation of motion, it is essential to completely understand the components of this equation as well as the system's working process. This book describes all possible components of the differential equation of motion and all possible factors of the working process. In mechanical engineering, all these components and factors represent forces and moments. The characteristics of all these loading factors and their application to particular differential equations of motion are presented in this book.

This book also introduces a straightforward universal methodology for solving differential equations of motion by using the Laplace Transform. This approach replaces calculus with conventional algebraic procedures that do not represent any difficulties for engineers. Using the Laplace methodology to solve differential equations of motion does not require memorizing the fundamentals of the Laplace Transform. Instead, this book presents an appropriate table of Laplace Transform pairs. It then explains how to use the pairs to convert differential equations into algebraic equations and then how to invert the solutions of these algebraic equations into conventional equations representing the functions of displacement of time.

Analyzing the solutions of differential equations of motion reveals the role of the system's parameters, the influence of these parameters on each other, and how to control the performance of the system.

The motion of a mechanical system is characterized by its displacement, velocity, and acceleration. These three characteristics are the three basic parameters of the system's motion. All other characteristics of the working processes can be determined by analyzing these three parameters. The equation of motion represents the displacement of the system as a function of time. The other two parameters — velocity and acceleration — are respectively the first and second derivatives from the displacement. Thus, the equation of motion is the basis for solving the mechanical engineering problem.

The equations of motion represent the solutions of differential equations of motion that reflect the real working processes of the systems. When we assemble these differential equations of motion, we use methodologies that are built on a close interaction between theoretical and applied sciences. Rapidly advancing technology stimulates intensive searches for more sophisticated engineering solutions. Therefore, we must be familiar with the methodologies for solving actual mechanical engineering problems.

This text can help you achieve the level of competence you need to successfully analyze real mechanical systems. An engineering educational background is sufficient to comprehend the contents of this text. We develop a comprehensive, step-by-step guide to solving mechanical engineering problems. Numerous examples demonstrate the methodologies that enable us to control the parameters of real systems. A wide range of readers can benefit from this book. Accounting for the different levels of their backgrounds, the step-by-step approach begins with the simplest examples and then gradually increases the complexity of the problems.

The text consists of six chapters. Let us consider briefly the contents of each.

## 1. Differential Equations of Motion

Our analysis of problems associated with dynamics is based on the laws of motion. These laws (or equations) of motion are the subject of Chapter 1. They represent displacement (the dependent variable, the function) as a function of running time (the independent variable, the argument). In general, motion has three phases: acceleration, uniform motion, and deceleration.

Displacement in uniform motion is a product of multiplying a constant velocity by the running time. This formula is known from basic physics; it is applicable to any uniformly moving object. Analysis of this formula, however, adds very limited help in understanding the working process and performance of an actual mechanical system.

Solutions that lead to performance control can be obtained from the expressions that describe acceleration and deceleration — equations representing the displacement, velocity, and acceleration as functions of time. For the plurality of real problems, there are no readily available formulas for these three parameters. Instead, mathematical expressions of these three parameters can often be obtained from solutions of corresponding differential equations of motion.

For each case, we should assemble an appropriate differential equation of motion that reflects the physical nature of the problem. As the book will show, composing differential equations of motion is not a trivial procedure.

A differential equation of motion is a second order differential equation made up of the second and first derivatives, the function, the argument, and, the constant terms. The structure of a second order differential equation is based on principles of mathematics without any dependence on laws of motion. The same approach is applicable to all mathematical rules used for practical calculations in different fields. The characteristics of motion include the second derivative (acceleration), the first derivative (velocity), the function (displacement), the argument (running time), and the constant terms. A natural linkage exists between the second order differential equation and the parameters of motion — the process of motion is described by a second order differential equation. (The second order differential equation is also applicable to electrical circuits and other physical phenomena; this text can be used for electrical engineering as well.)

In mathematics, the components of differential equations are dimensionless. In dynamics, each component of a differential equation should have the same physical units. Differential equations of motion are made up of loading factors that represent forces or moments whereas differential equations of electrical circuits include components that represent voltage.

The three basic parameters of motion are not loading factors — they have different units. These parameters cannot be directly included in a differential equation of motion. Each parameter should be multiplied by appropriate coefficients in such a way that the products have the units of loading factors, which cause the motion of objects.

Both the structure and the solution of the differential equations of motion are absolutely identical for rectilinear and rotational motions; their parameters are completely similar. Thus, the examples are presented just for rectilinear motion. Keep in mind that, if necessary, forces should be replaced by moments while the masses should be replaced by moments of inertia; the rectilinear parameters of motion should be replaced by the corresponding angular parameters. All this will not change the structure of the differential equation of motion and its solution. All considerations regarding forces are completely applicable to moments.

Particular attention is paid in Chapter 1 to explaining the structure of differential equations of motion and assembling them.

## 2. Analysis of Forces

The structure of the differential equation of motion is absolutely similar for rectilinear and rotational motion. So too is the process of composing the equation. To avoid redundant explanations, our analysis of loading factors focuses just on forces. However, the same characteristics and considerations are completely applicable to moments.

The left side of the differential equation of motion consists of resisting forces, whereas the right side consists of active forces. The resisting forces are variables (inertia, damping, and stiffness) and constants (e.g., dry friction, gravity, and plastic deformation). The force of inertia is present in all differential equations of motion. The resisting forces should be identified depending on the functionality and on the structure of the mechanical system as well as on the nature of the environment in which the motion occurs.

As variables, the force of inertia depends on acceleration, the damping force on velocity, and the stiffness force on displacement. These resisting forces can be linear or non-linear and their characteristics are determined by their coefficients. The coefficient of the inertia force is the mass, which is usually a constant value; consequently, the inertia force is linear. Non-linear inertia forces are not considered in this book. The damping and stiffness coefficients can also be constant or variable. If constant, the differential equation of motion is linear. If even one of these coefficients is a variable, the differential equation of motion is non-linear.

In certain mechanical systems, resisting forces could appear that represent some functions of time. However the majority of conventional mechanical systems do not have any obvious factors pointing to the existence of time-depending resisting forces which, therefore, are not discussed in this book.

In the majority of cases, the characteristics of active forces are predetermined. For conventional mechanical systems, these active forces include: constant forces, sinusoidal forces exerted by vibrators, and forces depending on time, velocity, or displacement. These last three can be linear or non-linear.

Chapter 2 looks closely at the characteristics and peculiarities of the resisting and active forces.

## 3. Solving Differential Equations of Motion Using Laplace Transforms

In solving the differential equations of motion, our goal is to obtain an expression for displacement as a function of time. This expression is also called the law of motion. Finding the best method for solving various linear differential equations can be challenging. However, the Laplace Transform represents a straightforward universal method for solving all linear differential equations.

The Laplace Transform lets us convert differential equations into algebraic equations whose solutions can be achieved by conventional algebraic procedures. We can apply the Laplace Transform without addressing the mathematical principles on which it is built. It provides a straightforward methodology regardless of the characteristics of the equation's components or its initial conditions.

Chapter 3 reviews the steps of this methodology; they are identical for each differential equation. First, we convert the differential equation of motion from the time domain form into the Laplace domain form, working with a table of Laplace Transform conversion pairs compiled for this text. The second step of the methodology deals with the Laplace domain solution of the differential equation of motion. This step, based on ordinary algebraic procedures, results in an algebraic equation that represents the dependant variable (e.g., displacement) as a function of the independent variable (e.g., running time). Both variables are in the same Laplace domain. The Laplace Transform eliminates the need of calculus to solve the differential equation of motion. Therefore, we obtain an algebraic equation with the dependent variable in the left side of the equation, and a sum of algebraic expressions (proper fractions) on the right.

In the last step, we invert all the terms of the solution from the Laplace domain into the time domain form. This inversion represents the solution of the differential equation of motion. All three steps of this methodology are demonstrated in the text by solving numerous examples.

In some cases, there will be terms in the right side of the Laplace domain solution that do not have representations in this text's table, or even in other, more comprehensive tables. For these cases, Chapter 3 discusses a method of decomposition used to resolving these expressions.

The examples in this chapter begin with a solution of a very simple differential equation. The complexity of the solutions gradually increases; ultimately, the examples include a range of diversified differential equations of motion of actual mechanical systems.

## 4. Analysis of Typical Mechanical Engineering Systems

Assembling the different equation of motion is a very important step when investigating the dynamics of a mechanical system. The differential equation should reflect the peculiarities of the real working process. This chapter discusses the considerations that are relevant to the process of assembling differential equations of motion. These considerations are associated with real-life problems of typical mechanical systems. We start with composing the appropriate differential equation of motion. The following step focuses on this equation's solution. In the last step, our analysis of the solution reveals the system's performance characteristics: energy consumption, required power, acting forces, and others. The complexity of the examples increases from example to example, and can be very helpful in solving actual problems.

## 5. Piece-Wise Linear Approximation

Chapters 3 and 4 are devoted to solving linear differential equations of motion. In reality, many loading factors that are included in these equations are actually non-linear. However, the non-linearity of these factors is often not essential; it is then justifiable to consider them as being linear. There are no currently established methodologies for solving non-linear differential equations in general terms.

Many specific non-linear differential equations can be solved using particular mathematical investigations, and there are catalogs where these solutions can be found. However, these solutions have a very limited applicability to non-linear differential equations of motion.

In a significant number of real life problems, the non-linearity of the loading factors cannot be ignored. Neglecting the strong non-linearity of these factors results in essential quantitative errors; yet some important qualitative characteristics of the process could be misunderstood or not revealed at all.

The method of piece-wise linear approximation allows us — with an appropriate accuracy — to investigate problems that include non-linear loading factors. The characteristics of these factors can be represented by corresponding graphs whose curvatures reflect the extent of the factors' non-linearity.

Piece-wise linear approximation consists of replacing the curve by a broken line. For instance, if the curve is replaced by a broken line including three straight segments, the process of motion can be divided into three intervals. For each interval, a linear differential equation will be composed with the initial conditions of motion equal to the conditions of motion at the end of the previous interval. The shorter the length of the segments, the more accurate the results of the solution will be. A reasonable compromise will decide the number and values of the replacement increments that would satisfy the goal of the investigation. The application of the piece-wise linear approximation to the solutions of real-life problems comprising non-linear loading factors is presented in a detailed way in this chapter.

## 6. Dynamics of Two-Degree-of-Freedom Systems

Numerous mechanical engineering systems are made up by several separate masses connected among themselves by specific links. These links allow for motion of these masses relative to each other. Each motion is described by its mass's differential equation. The amount of these masses defines the number of degrees-of-freedom of the system. Of the actual multiple-degree-of-freedom mechanical systems, the majority have just two masses — therefore, this text is limited to considering two-degree-of-freedom structures. Two types of links allow relative motion of the connected masses: the elastic link (spring) and the hydraulic link (dashpot). The masses could be connected by a hydraulic or elastic link, or by both links acting in parallel. A simultaneous system of two differential equations of motion should be assembled in order to describe the motion of the two masses.

Chapter 6 contains a detailed discussion of the structures of the differential equations of motion and also of the considerations for composing these equations. It also includes typical examples that demonstrate the methods for investigating two-degree-of-freedom systems.

## A General Note

These chapter descriptions indicate that the analysis of an actual mechanical system is a complex process engaging an interaction among several sciences.

During the first steps of the analysis, we should pay particular attention to the characteristics of the damping and stiffness resisting forces. In the majority of practical cases, these forces could be linear or non-linear whereas the rest of the forces are usually linear. Information regarding the characteristics of the actual damping and stiffness forces for a specific case should be based on the results of the investigations; these results are usually presented in graphs or can be found in corresponding sources.

Normally, our analysis of the solutions of the differential equation of motion provides the information needed to make appropriate engineering decisions. This text includes all the steps necessary for a complete analysis of actual problems in mechanical engineering dynamics.

Numerous software programs are available for computing the parameters of motion of mechanical engineering systems. These programs can be used when the differential equations of motion are already available. When investigating real life problems, the first steps are associated with composing the differential equations of motion. This text is intended to help you assemble these equations. In many practical situations, you may need to analyze the working process of a mechanical engineering system in order to estimate the influence of the parameters on each other and to reveal their specific roles. For these cases, we present the analysis in general terms without any use of related numerical data. This book will also be useful for performing this kind of analyses.

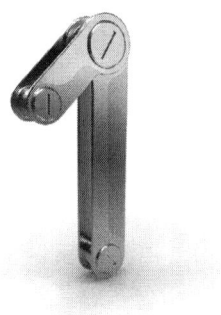

# DIFFERENTIAL
# EQUATIONS OF MOTION

A mechanical system's equation of motion, also called the *law of motion*, represents the system's displacement as a function of running time. Analyzing the equation of motion provides comprehensive information needed for the development, design, and improvement of the system. The equation of motion is the solution of the differential equation of motion for the system performing a certain working process.

The accuracy of the analysis can be evaluated by appropriate experiments. Results that disagree with the experiments tell us the differential equation does not closely enough reflect the actual conditions of the process of motion. In such cases, we revise the differential equation. We may need to carry out a few iterations to achieve the acceptable accuracy; however, in many practical cases, our first iteration should be enough. The considerations presented below may help us develop these equations.

## 1.1   The Left Side of Differential Equations of Motion (Sum of Resisting Loading Factors Equals Zero)

From Dynamics, we know that the differential equation of motion is a second order differential equation. As it turns out, a second order differential equation also describes the processes in electrical circuits. The structure of such an equation is predetermined by principles of mathematics without any regard to either the characteristics of motion of a mechanical system or the characteristics of processes in electrical circuits. An ordinary linear second order differential equation reads:

$$c_1 \frac{d^2 x}{dt^2} + c_2 \frac{dx}{dt} + c_3 x + P = f(t) \tag{1.0}$$

where $x$ is the function, $t$ is the argument, $c_1, c_2$, and $c_3$ are constant coefficients, $P$ is a constant value, and $f(t)$ is a certain known function of $t$.

Let's examine the left side of this equation. The first term is the product of multiplying a constant coefficient by the second derivative. The second term is the product of multiplying another constant coefficient by the first derivative. The third term is the product of multiplying one more constant coefficient by the function, and finally the last term is a constant value. This constant value can be considered a product of multiplying a constant coefficient by the function (or argument) to the zeroth power.

The right side of this equation may include certain known variable and constant values. All these terms must either have the same units or be dimensionless. The solution of equation (1.0) represents an expression describing the dependence between the function $x$ and the argument $t$.

Now let's consider equation (1.0) from the viewpoint of Dynamics. Displacement, velocity, and acceleration are the three basic parameters of motion of a mechanical system. All other parameters are derived from these three. Hence, the left side of a second order differential equation helps describe the motion of a system because it contains the same basic structural parameters: the second derivative (acceleration), the first derivative (velocity), the function (displacement), and a constant value. Newton's Second Law states that a body's motion is caused by a force. This Second Law is expressed by the following well known formula:

$$F_0 = m_0 a_0$$

where $F_0$ is the force, $m_0$ is the mass of the body, and $a_0$ is the acceleration of the motion of the body (the indexes "0" are given in order to avoid possible confusion with similar parameters in the text).

The first term of equation (1.0) contains the second derivative, which is the acceleration. According to Newton's Law, the coefficient $c_1$ in the differential equation of motion should be replaced by the mass $m$. Thus, the first term of the differential equation of motion is actually a force; all other terms of this equation should have the same units and they should be forces. The product of multiplying the mass by the acceleration (second derivative) represents the force of inertia. Because the mass is a constant value and in general the second derivative (the acceleration) is a variable quantity, we conclude that the force of inertia depends on the acceleration. Similarly, by multiplying the second and third terms of equation (1.0) by certain specific coefficients, we obtain respectively a force that depends on the velocity and a force that depends on the displacement.

The force that depends on the velocity is actually the reaction of a fluid medium to a movable body that interacts with this medium. This reaction represents a resisting damping force; the coefficient at the first derivative (the velocity) is called the *damping coefficient*.

The damping coefficient depends on both the type and condition of the fluid and also on the shape and dimensions of the movable object. Special hydraulic links (dashpots) are used in some mechanical systems in order to absorb impulsive loading. These links exert damping forces and are characterized by damping coefficients. Very often the damping coefficient depends on the velocity of the movable body.

When this coefficient does not depend on the velocity, or the dependence is negligible, the damping coefficient is considered to have a constant value. In this case, the differential equation of motion is linear — assuming, of course, that all other components of the equation are linear. If instead this coefficient has a variable value, the equation becomes non-linear.

There are no readily available formulas to calculate the damping coefficient. For each case, the characteristics and the value of the damping coefficient should be determined on the basis of experimental data. Note that in some cases a damping force becomes a part of the external loading factors (see Chapter 4).

The forces that depend on the function (the displacement) are exerted by the elastic media in the process of interacting with a movable object. By its nature, this force is the reaction of the medium to its deformation by a movable body. This force represents a resisting force and is called the *stiffness force*. The function's coefficient is the *stiffness coefficient;* it depends on the type and condition of the elastic medium, the shape and dimensions of the body, and the peculiarities of the deformation.

For some elastic media, the stiffness coefficient depends on displacement of the movable object (deformer). If this dependence is negligible, the stiffness coefficient is considered to have a constant value. The corresponding differential equation of motion is linear. If, however, this dependence is significant, the stiffness coefficient is characterized by a variable value. The corresponding equation of motion is then non-linear.

Mechanical engineering systems often include elastic links in the shape of springs. The stiffness coefficient for the springs can be calculated using readily available formulas. Sometimes this coefficient is called the *spring constant*. For deformation of elastic media, there are no readily available formulas to calculate the stiffness coefficients; appropriate data is needed to determine the values and characteristics of these coefficients. In some cases, the stiffness force is considered an external active force (see Chapter 4).

The fourth term of equation (1.0) has a constant value. In the differential equation of motion, this value may be represented by certain constant resisting forces such as the force exerted during the deformation of a plastic medium, the dry friction force, or the force of gravity in case of an upward motion.

Consider the right side of equation (1.0) with respect to the differential equation of motion. This part may comprise a force that is a certain known function of time, velocity, or displacement — or a sum of all of them, including a constant force. In a very specific case (Chapter 4), the right side of the differential equation of motion contains a force that depends on acceleration. All these considerations let us conclude that the structure of a differential equation of motion is determined by equation (1.0).

With respect to Dynamics, the terms in the left side of equation (1.0) represent forces that resist the motion of a mechanical system, whereas the right side of the equation includes terms that cause the motion. The forces that resist the motion characterize the reaction of the system to its motion. Thus, the forces that have a reactive nature are the resisting forces. The forces in the right side of the differential equation of motion are applied to the system; they may be called the *external forces* or the *active forces*.

More considerations associated with loading factors (forces and moments) and with the structure of the differential equations of motion are discussed below.

Like any equation, a differential equation of motion consists of two equal parts. The components of the equation represent forces or moments applied to the mechanical system. Forces are used in equations of a particle's rectilinear motion or a rigid body's rectilinear translation, whereas moments are used for equations to describe the rotation of a body around its axis. The forces or moments can be classified into two groups:

1. Active forces and moments causing motion
2. Reactive forces and moments resisting the motion

It is justifiable to place all the resisting loading factors into the left side of the differential equation, and the active loading factors into the right side.

Based on all these considerations, it is possible to assemble the most general left side for an actual mechanical engineering system's differential equation of motion. Let's start with a system in the rectilinear motion. Assume that no external forces are applied to the system, which is moving in a horizontal direction. In this case, the right side of the equation equals zero. In the absence of external forces, the motion occurs due to the energy that the system possesses (kinetic, potential, or both). The initial conditions of motion contain information regarding the energy the system possesses at the beginning of the analysis. Based on these considerations and applying equation (1.0), we may write the differential equation of motion of a mechanical system, as seen in equation (1.1):

$$m\frac{d^2x}{dt^2} + C\frac{dx}{dt} + Kx + P + F = 0 \qquad \textbf{(1.1)}$$

where $x$ is the displacement (the function), $t$ is the running time (the argument), $m$ is the mass of the system, $C$ is the damping coefficient, $K$ is the stiffness coefficient, $P$ is the constant resisting force of any nature except friction, and $F$ is the constant dry friction force. The force $P$ could represent the force of the system's gravity in cases of upward motion, the resisting force exerted during a medium's plastic deformation, or other. The gravity force becomes an active force when it acts in the direction of motion.

The first term of equation (1.1) is the force of inertia; it represents the product of multiplying mass $m$ by acceleration $\frac{d^2x}{dt^2}$. The second term of this equation is the resisting damping force; it depends on the system's velocity. This force equals the product of multiplying the damping coefficient $C$ by the velocity $\frac{dx}{dt}$. The nature of the damping force is related to fluid (liquid and air) resistance, which represents the reaction of a viscous medium during its interaction with a movable object.

The third term of this equation is the stiffness resisting force that depends on the system's displacement. This force equals the product of multiplying the stiffness coefficient by the displacement $x$. The nature of this force is the reaction of an elastic medium to its deformation by a movable body. By its nature, the resisting constant force associated with a medium's plastic deformation and the dry friction constant force are also reactive forces. More details about the force of inertia, the damping and stiffness forces, and their coefficients are presented below.

From a mathematical point of view, assume that the left side of equation (1.1) could include a resisting force dependant on time. As a matter of fact, all forces in the differential equation of motion are functions of time, including the constant forces that are actually coefficients at the time that is to the zeroth power.

Let's analyze a hypothetical case where a resisting force depends directly on time. Imagine a device programmed to increase a pressure force dependant on time. This device, which is attached to a movable system, applies a resisting force that is increasing in time. But this force is not reactive by nature. Instead, it is an external or active force; it should be included in the right side of the differential equation. Thus, a force that depends directly on time should not be included in the left side of a differential equation of motion. All this makes it clear that equation (1.1) represents the most general left side of a differential equation of motion that includes all possible resisting forces; it describes the rectilinear motion of a hypothetical mechanical system in the absence of external forces.

In order to solve the differential equation of motion for this system, first determine the initial conditions of motion:

$$\text{For} \quad t = 0 \quad x = s_0, \quad \frac{dx}{dt} = v_0 \tag{1.2}$$

where $s_0$ and $v_0$ are the initial displacement and velocity respectively.

According to these initial conditions, the system possesses the potential energy of the deformed medium — the deformation is proportional to the initial displacement $s_0$. The system also possesses the kinetic energy that is proportional to the initial velocity $v_0$. In this case, the system's motion is caused by the combined action of potential and kinetic energy.

Equation (1.1) describes the motion in cases where the system possesses just the potential energy (for $t = 0$ $x = s_o$, $\frac{dx}{dt} = 0$), or just the kinetic energy (for $t = 0$ $x = 0$, $\frac{dx}{dt} = v_0$). For each of these cases, however, the solutions of the differential equation (1.1) will be different. (This will be demonstrated in Chapter 3.) When both the initial displacement and the initial velocity equal zero, there will be no motion.

The left side of equation (1.1) has five components or, in the general case of the differential equation of motion, these five resisting forces:

1. Force of inertia $m \frac{d^2t}{dt^2}$
2. Damping force $C \frac{dx}{dt}$
3. Stiffness force $Kx$
4. Constant force $P$
5. Dry friction force $F$

These forces represent the reaction of all possible factors to the system's motion. Their characteristics depend on the structure of the mechanical system and on the nature of the environment in which the system is moving.

The structure of the left side of the linear differential equation of motion (1.1) corresponds to the structure of the second order linear differential equation. Not all resisting forces are present in each actual problem; in reality, the left side of the equation may include any number of components from one to five. However, the force of inertia associated with the second order derivative is always present in the differential equation of motion. Thus, the shortest and simplest expression of a differential equation of motion is that the force of inertia equals zero, and the motion is caused by the initial velocity. In this case, the body moves by inertia with a constant velocity (the acceleration equals zero). This case is discussed further in Chapter 3.

Now let's compose a similar differential equation of motion for a body in rotation around its horizontal axis. We apply the same procedures as before:

$$ J \frac{d^2\theta}{dt^2} + C \frac{d\theta}{dt} + K\theta + M_P + M_F = 0 \qquad \textbf{(1.1a)} $$

The initial conditions are:

$$\text{For} \quad t = 0 \quad \frac{d\theta}{dt} = \Omega_0, \ \theta = \theta_0 \qquad \textbf{(1.2a)}$$

where $J$ is the moment of inertia of the system, $C$ and $K$ are respectively the damping and stiffness coefficients in rotation, $M_P$ is a resisting constant moment, $M_F$ is a constant moment associated with dry friction, $\theta$ is the angular displacement, $t$ is the running time, and $\Omega_0$ and $\theta_0$ are the initial angular velocity and initial angular displacement respectively. As indicated above, equations (1.1) and (1.1a) are absolutely similar as are expressions (1.2) and (1.2a). The solutions of these equations with their initial conditions are also absolutely similar.

## 1.2 The Left and Right Sides of Differential Equations of Motion (Sum of Resisting Loading Factors Equals Sum of Active Loading Factors)

The left side of the differential equation of motion comprises the resisting forces or moments whereas the right side consists of active or external forces or moments. These two parts are equal to each other.

Now let's compose the differential equation of the rectilinear motion for a body that is subjected to all possible resisting and active forces. The left side of this equation is the same as in equation (1.1). Thus, we must compose the right side so that it includes all possible active or external forces. According to the structure of the second order differential equation, its components could represent

1. Constant values
2. Variables depending on the argument
3. Variables depending on the function
4. Variables depending on the derivatives of the function

In other words, the active forces include constant and variable forces. The variable forces may depend on the time, the displacement, the velocity, and the acceleration.

As an example of where the active force depends on the acceleration, consider the motion of a rear-wheel or front-wheel drive automobile. The automobile's force of inertia is responsible for the redistribution of its force of gravity between the rear and front axels. This redistribution causes a certain increase of the rear axle's loading, and a corresponding decrease of loading on the front axle.

For a rear-wheel drive automobile, the increase of loading on the rear axle leads to the increase of the friction force between the rear wheels and the ground. This results in the increase of the active force that causes the motion of the automobile. For a front-wheel drive automobile, the force of inertia plays an opposite role — the active force decreases. An example is provided in Chapter 4.

In this example, the value of the force of inertia from the right side of the automobile's differential equation of motion represents just a small fraction of the total force of inertia from the left side of the equation. In the case of a four-wheel drive automobile, the redistribution of the weight between the axles does not influence the total resultant active force.

It is problematic to find other examples where an active force depends on the acceleration. Furthermore, in the rotational motion, it would probably be impossible to find a situation where the angular acceleration influences the applied active moment. Therefore, it is justifiable not to include in the right side of the equation a force that depends on acceleration.

Figure 1.1 shows a general case of a variable active force that is dependant on time. This graph approximates a random variable force. It represents the action of a random force by using a sinusoidal curve that has a maximum force $R_{max}$, and a minimum force $R_{min.}$

The mean force $R$ divides the graph into two equal parts; it is calculated from the following formula:

$$R = \frac{R_{max} + R_{min}}{2} \qquad (1.3)$$

Force

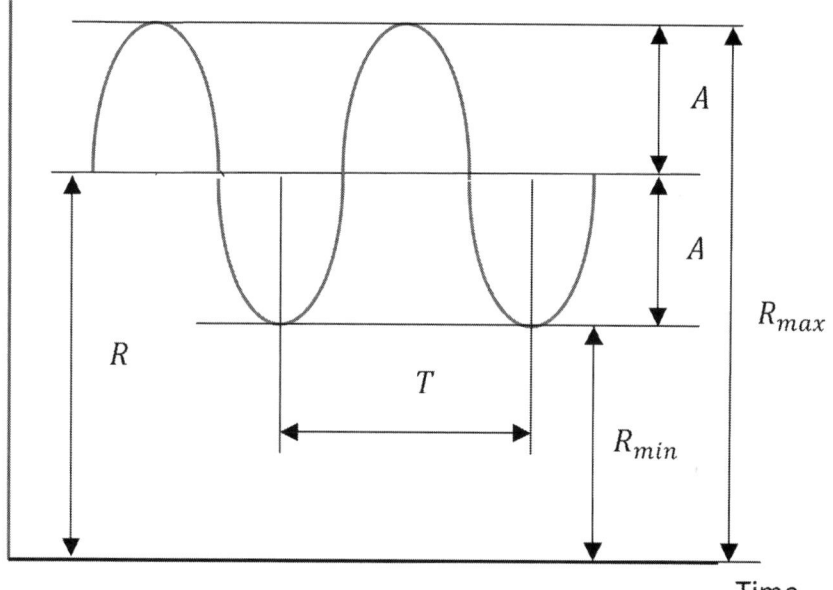

**Figure 1.1  The variable random active force.**

The amplitude $A$ of the sinusoidal force can be determined from equation (1.4):

$$A = \frac{R_{max} - R_{min}}{2} \qquad (1.4)$$

The frequency $\omega_1$ of the sinusoidal force is:

$$\omega_1 = \frac{1}{T} \qquad (1.5)$$

where $T$ is the period of fluctuation of the sinusoidal force.

A variable random force can be replaced by a superposition of a constant force $R$ and a sinusoidal force $A \sin \omega_1 t$. The sinusoidal force is a harmonic function of time.

Active forces can be expressed as linear or non-linear functions of time. In most practical cases, these active forces are considered as linear functions of time. Active forces can also be presented as functions that are dependent on displacement or velocity. Based on all these considerations, we assemble equation (1.6), the most general differential equation of motion of a mechanical system that moves in the horizontal direction:

$$m\frac{d^2x}{dt^2} + C\frac{dx}{dt} + Kx + P + F$$

$$= R + A\sin\omega_1 t + Q\left(1 + \frac{\mu t}{\tau}\right) + C_1\frac{dx}{dt} + K_1 x \qquad (1.6)$$

where $R$ is a constant active force; $A$ is the amplitude of a sinusoidal force; $\omega_1$ is the frequency of the sinusoidal force; $Q$ is the constant value of an active force at the beginning of the motion; $\tau$ is the time that the motion can last; and $\mu$ is a constant dimensionless coefficient, for $\tau > 0$, $\mu > 0$, and $t \leq \tau$; and finally $C_1$ and $K_1$ are respectively the damping and stiffness coefficients of the active damping and stiffness forces.

The initial conditions for this case are arbitrary, which means that the parameters of motion at the beginning of the process could be equal to zero or could be different from zero. In general, the initial conditions of motion for equation (1.6) may be also presented by the expression (1.2).

The left side of equations (1.1) and (1.6) are identical, as expected. Thus, equation (1.6) represents the structure of the most general differential equation of a rigid body's rectilinear motion. The complexity of the differential equations of motion in the actual situations is significantly lower than in equation (1.6).

The characteristics of a differential equation of motion's forces determine the linearity or non-linearity of the equation.

It is very important for the overall analysis to clarify the peculiarities of the characteristics of the forces that should be included in the equation. There are two main concerns in determining the characteristics of the forces.

1. The first is associated with obtaining the most credible data about the particular forces for the particular real conditions.
2. The second is related to interpreting these characteristics in terms of their linearity or non-linearity.

Adequate information is obtainable from a comprehensive search of the relevant sources. The decision to categorize these characteristics as linear or non-linear depends on both the actual experimental data and the level of compromise that is justifiable in each particular case. The following analysis of these forces addresses these two concerns.

These considerations are all related to those differential equations of motion in the horizontal direction that are not affected by vertical forces such as gravity. In cases of vertical motion or an incline, the force of gravity plays a role. If a mechanical system is moving up vertically or on an incline, the force of gravity or its component represents a resistance and should be included in the left side of the differential equation of motion. When the system moves down, these forces represent active or external forces and should be included in the right side of the equation. According to equation (1.6), the right side of the differential equation of motion generally includes five active forces:

1. Constant force $R$
2. Sinusoidal force $A \sin \omega_1 t$
3. Force depending on time $Q\left(1 + \frac{\mu t}{\tau}\right)$
4. Force depending on velocity $C_1 \frac{dx}{dt}$
5. Force depending on displacement $K_1 x$

Detailed descriptions of the characteristic of forces included in the differential equations of motion (1.1) and (1.6) are presented in Chapter 2.

Now let us compose a differential equation of motion for a system that rotates around its horizontal axis and is subjected to all possible resisting and active moments. This equation is completely similar to equation (1.6) and is presented in the following shape:

$$J\frac{d^2\theta}{dt^2} + C\frac{d\theta}{dt} + K\theta + M_P + M_F$$

$$= M + M_A \sin\omega_1 t + M_Q\left(1 + \frac{\mu t}{\tau}\right) + C_1\frac{d\theta}{dt} + K_1\theta \quad \textbf{(1.6a)}$$

where $M$ is a constant active moment; $M_A$ is the amplitude of a sinusoidal active moment; $\omega_1$ is the frequency of the sinusoidal moment; $M_Q$ is the constant value of an active moment at the beginning of the motion; and $C_1$ and $K_1$ are respectively the damping and stiffness coefficients; $\mu$ is a constant dimensionless coefficient, and $\tau > 0$, $\mu > 0$, and $t \leq \tau$.

The initial conditions of motion for equation (1.6a) are presented in expression (1.2a).

Because of the strong similarity between the characteristics of forces and moments, as well as between the appropriate differential equations of motion, we will consider only forces in this text while keeping in mind that all considerations related to forces are also applicable to moments.

It is important to realize that all components in the differential equation of motion should be functions of time. (For constant terms, time is to the zeroth power.) In certain real situations, the movable mechanical systems are subjected to forces that may depend on temperature. Changes of fluid temperature will cause changes of the damping force. The influence of temperature on forces cannot be directly incorporated into the differential equations of motion. If we could also incorporate forces that depend on temperature into these equations, we would obtain equations with two independent variables: time and temperature. There are no differential equations of motion with multiple arguments; there can be just one independent variable — and it must be the running time. Therefore, any forces that depend on temperature or other factors, except time, cannot be included in any differential equations of motion.

However, there is a way to account for the change of the forces due to temperature change. Consider the change of fluid viscosity due to temperature. As mentioned above, this change will cause the change of the damping force. In other words, temperature change will result in the change of the damping coefficient. The differential equation of motion and its solution are the same for different values of damping coefficients. These values should be determined for different temperatures and then used during the quantitative analysis of the parameters of motion. The results of this analysis will reveal the influence of temperature on the process of the system's motion.

# ANALYSIS OF FORCES

The differential equations of motion are made up of resisting and active forces. The resisting forces are also called the *reactive forces* or the *resistance forces*. The active forces are sometimes called the *external forces*.

## 2.1 Analysis of Resisting Forces

Resisting forces include the forces of inertia, damping forces, stiffness forces, constant forces, and friction constant forces. All of these forces are products of multiplying certain coefficients by certain functions of running time. The constant forces are the coefficients, with the running time set to the zeroth power.

The analysis of these forces focuses on the peculiarities of their coefficients. If these coefficients have constant values, the differential equations of motion are linear and can be solved in general terms. If, however, even one of these coefficients is a variable quantity, the equation becomes non-linear; in the majority of situations, they then cannot be solved in general terms. In reality, these coefficients are often non-linear. Sometimes the non-linearity is not essential and can be ignored. But in many cases, the non-linearity is strong and should be taken into consideration.

The appropriate experimental data reveals the characteristics of these coefficients. In most cases, they are represented by curves. There are no definite criteria that indicate what curvature is the border between linear and non-linear parameters. Your decision should be based on the many factors associated with the problem. If non-linearity is essential, a curve may undergo linearization by using piece-wise linear approximation. According to this methodology, the curve is replaced by a corresponding broken line; linear differential equations are then composed and solved for each segment of the broken line. You determine the length and number of the segments. Shorter segments result in more accurate solutions, but require more work. This methodology is considered further in the following sections.

### 2.1.1  Forces of Inertia

The force of inertia is a product of multiplying the mass of the system by the second derivative of the function, which represents the acceleration or deceleration of the mass. The mass represents the coefficient that should be analyzed to determine if its value is constant or variable. In general, the majority of mechanical systems have constant masses. In certain conditions, however, the masses may change their values due to factors such as evaporation, melting, rain, and snow. Transportation systems powered by internal combustion engines have decreasing masses due to fuel consumption.

When the mass depends on acceleration, the structure of the force of inertia remains unchanged, namely $m\frac{d^2x}{dt^2}$, where $m$ could be a constant or variable value. In other cases the change of the mass may depend on time, velocity, displacement, temperature, other factors, or a combination of factors. In cases involving fuel consumption, assume that — along with other factors — the decrease of the mass has a certain relationship with the acceleration. There are no mathematical expressions that could be included in the differential equation of motion to account for these changes of mass.

Analyzing experimental data (often expressed in graphs) helps us to determine the characteristics of the coefficients involved in differential equations of motion. If a graph has a straight line, the coefficient has a constant value whereas if the graph is curved, the coefficient has a variable value. It is difficult to come up with an actual example in which the graph representing the force of inertia's dependence on acceleration is curved. The solution of a differential equation with a variable mass is a complicated procedure and is not included in this text.

The force of inertia plays a specific role in wheeled ground transportation; it causes a redistribution of the vehicle's weight between its axles. This redistribution has a certain influence on the dynamics of an automobile. More details are discussed in Chapter 4.

## 2.1.2 Damping Forces

The motion of a body in a surrounding fluid medium (air, or water, or both) causes a deformation of that medium. The resistance of the fluid medium to its deformation is expressed by damping forces applied to the movable body. These forces represent the reaction of the fluid medium to its deformation by the movable body. The distinctive characteristic of the resisting damping forces is that they depend on the velocity of their deformation. A damping force equals the product of multiplying the damping coefficient by the velocity of the movable body. The damping coefficient depends on both the fluid viscosity and the shape and dimensions of the body. In some fluids, this coefficient may also depend on the velocity of the body. If the damping coefficient has a constant value, the differential equation is linear. In actual conditions, this coefficient may change its value depending on the relative velocity between the body and the medium. In such cases, the differential equation becomes non-linear.

First obtain the appropriate data; you can usually use graphs to reflect the relationship between the damping force and velocity. Analyzing these graphs helps you understand the characteristics of the damping coefficients. If the graph represents a straight line the coefficient has a constant value. If instead the graph has a curved shape, the coefficient has a variable value. Piece-wise linear approximation allows us to analyze variable damping coefficients. The details of this procedure are shown in the following sections.

Specific hydraulic links are used in many mechanical engineering applications to absorb impulsive loading by developing resisting damping forces. These links, which embody all kinds of shock absorbers and vibration-damping devices, are generally called *dashpots*. A dashpot represents a piston-cylinder system filled with fluid (see Figure 2.1). The movable mass 1 is rigidly connected to the piston 2, which moves inside of a cylinder 3, forcing the fluid 4 to flow through calibrated orifices (not shown). The cylinder 3 is securely connected to a non-movable support. The arrows in Figure 2.1 indicate the possible directions of the mass's motion.

There are no readily available formulas to calculate the damping coefficients. Their values should be obtained from appropriate experimental data.

You can use images of a dashpot, as seen in Figure 2.1, to symbolize mathematically or physically the presence of a damping resisting force in a movable system. Let's consider a few related

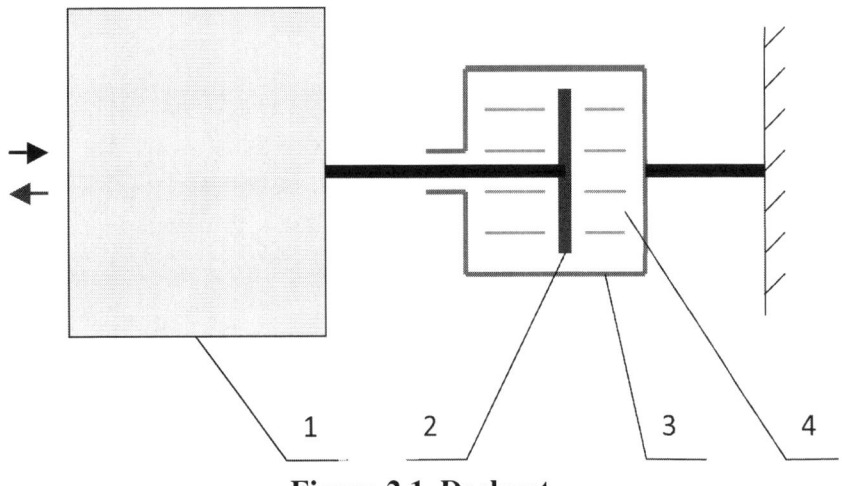

**Figure 2.1  Dashpot**

examples. Automobiles and aircraft are subjected to damping resistant forces exerted by the air. A submarine experiences damping action from the water. There are no physical dashpots attached to the automobile, the aircraft, or the submarine. However, in the mathematical models of these systems' motion, the attached dashpots are present; they symbolize the damping resisting forces acting in the opposite direction of the motion of these objects. The damping resistance of fluid media represents distributed forces applied to the surface of these systems. Yet in the mathematical models of motion of these objects, the distributed forces are replaced by concentrated damping forces applied to these objects (and presented by dashpots). These models are shown in the following sections.

Some mechanical systems, such as shock absorbing mechanisms, comprise real dashpots. In the mathematical models of these systems, the damping forces are also presented by dashpots.

In the examples considered above, the mechanical systems are subjected to the action of one resisting damping force. However, multiple damping resisting forces can be applied to a movable body. For example, a ship overcomes a combination of damping resistance from the water and from the air. For the ship, these two resistance forces are applied in parallel. In some shock absorbers, the damping forces may act in a sequence. In general, a body can be simultaneously subjected to the action of a combination of parallel and sequential damping devices.

As shown in Figure 2.2, the motion of the mass $m$ is affected by a system of six dashpots which are grouped in three parallel rows; each row has two dashpots in sequence. All dashpots are oriented in the same direction and each dashpot is characterized by its damping coefficient. The coefficients are $C_1$ and $C_2$ for the dashpots in sequence of the upper row, $C_3$ and $C_4$ in the middle row, and $C_5$ and $C_6$ in the lower row. In this case, the resulting damping coefficient $C$ equals the sum of all particular damping coefficients:

$$C = C_1 + C_2 + C_3 + C_4 + C_5 + C_6$$

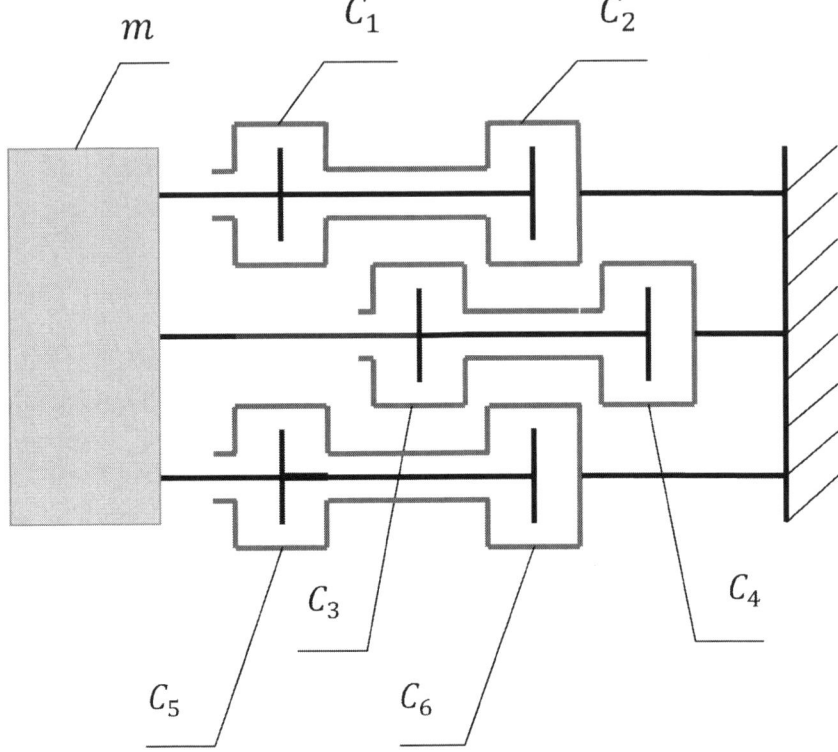

**Figure 2.2 A combination of parallel and sequential dashpots.**

In general, if the dashpots act in the same direction, the resulting damping coefficient represents the sum of each particular coefficient, regardless of the order in which the dashpots are attached to the mass. Based on these considerations, we can write the following expression for calculating the damping coefficient $C$:

$$C = \sum_{0}^{n} C_i \qquad (2.1)$$

where $i = 1, 2, 3, \ldots, n$, and $n$ is the number of dashpots.

In all mathematical models of motion, none of the imaginary and real dashpots have any masses. The dashpots simply represent the concentrated damping forces and indicate the direction of these forces.

### 2.1.3 Stiffness Forces

The nature of resisting stiffness forces is the reaction of an elastic medium to its deformation by a movable body. The distinctive characteristic of the resisting stiffness force is that they depend on the value, not the velocity, of the deformation. The value of the deformation is proportional to the displacement of the movable body.

Hammering a nail into wood, driving a pile into soil, and penetrating a bullet into steel are examples of deformation of different media. During the start of the deformation process of solid media, resisting forces increase as the deformer's displacement increases. This relationship reflects the elastic properties of the media. However, at a certain point of the deformer's displacement, the resisting force reaches its maximum value and then remains constant while the displacement continues. Thus, the deformation reaches the elastic limit and the deformer enters into the plastic stage of the medium's deformation.

Actually, the relationship between the resisting force and deformation of materials can be characterized by so called elasto-plastic or visco-elasto-plastic models of the media. There are no engineering materials that are just elastic or just plastic. The majority of these materials exhibit properties of both; some materials also possess viscosity.

There are no purely elastic materials. Elasticity is a property of engineering materials; usually the beginning of deformation is characterized by elastic behavior. Note that the stiffness force is exerted during the engineering materials' elastic stage of deformation. All kinds of vibrations of mechanical structures are associated with elastic deformations.

A stiffness force is a product of multiplying the medium's stiffness coefficient $K$ by the mechanical system's displacement $x$ (or position). The system's displacement is proportional to the medium's deformation. There are no readily available formulas to calculate the stiffness coefficients of the engineering materials; they depend on the type and conditions of the elastic media and on the shape and dimensions of the deformer.

In each particular case, the graphs representing the data can help us determine the stiffness coefficient. The stress-strain diagrams of different materials reveal the characteristics of their stiffness coefficients. If the diagram is in a form of a straight line, the stiffness coefficient has a constant value and the differential equation of motion is linear. If the diagram is in a form of a curved line, the stiffness coefficient has a variable value and the differential equation is non-linear. For the majority of materials, you can find stress-strain diagrams in numerous publications. If we have a variable stiffness coefficient, the method of piece-wise linear approximation could be applied. An appropriate example demonstrating this method is discussed in Chapter 5 of this book.

Many mechanical systems use specific elastic links. These links represent all kind of springs; they play different roles in the systems' working processes. The springs also exert stiffness forces during their deformation. A conventional vibratory system exemplifies the mechanics of the interaction between a movable body and an elastic link (spring). The model of such a system is shown in Figure 2.3, where the movable mass 1 is attached to an elastic link 2 representing a spring that is characterized by a stiffness coefficient $K$. The spring is secured to a non-movable support. In most situations associated with a spring, the stiffness force represents a resisting force. In a vibratory system, however, the spring plays the role of a resisting force during just half of the period of the vibration. A deformed spring accumulates potential energy of deformation; during the second half of the period the spring then releases this energy, resulting in the motion of the mass into the opposite direction.

The formulas to calculate the stiffness coefficients of springs are available in appropriate sources.

However, in cases of deformation of elastic media by specific rigid bodies, the data describing the characteristics of the corresponding stiffness coefficients can be found in appropriate sources. In some mechanical systems, the movable mass can be attached to a combination of elastic links acting in parallel or sequential order. Figure 2.4 shows a case where two springs act in parallel.

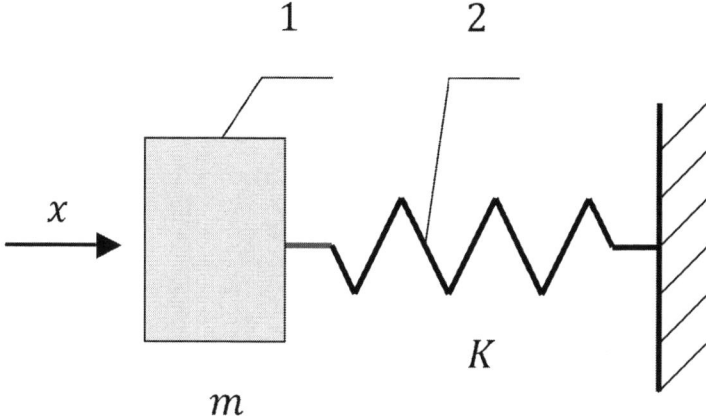

**Figure 2.3  Vibratory system.**

When the elastic links act in parallel, the stiffness coefficient $K$ equals the sum of all stiffness coefficients. For the case shown in Figure 2.4, we have:

$$K = K_1 + K_2$$

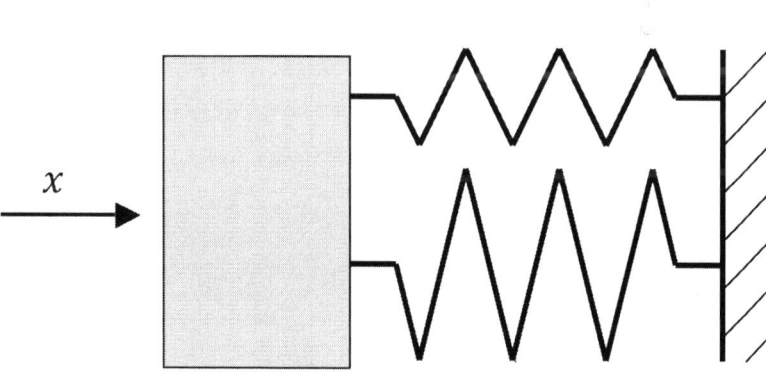

**Figure 2.4  Two springs act in parallel.**

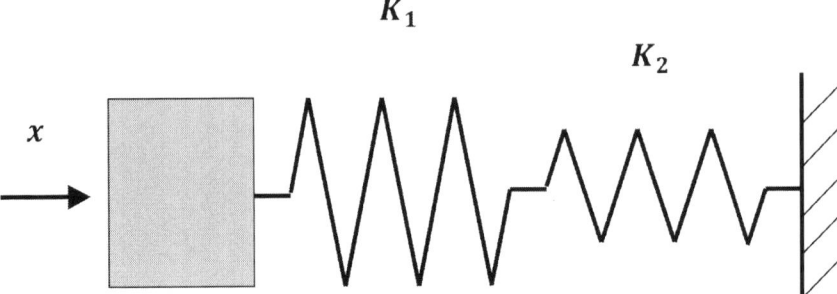

**Figure 2.5 Two springs act sequentially.**

In general, for a group of elastic links acting in parallel, the resulting stiffness coefficient $K_p$ equals:

$$K_p = \sum_{o}^{n} K_i \qquad (2.2)$$

where $i = 1, 2, 3, \ldots, n$, and $n$ is the number of springs.

Figure 2.5 shows a vibratory system with two springs that are sequentially attached to the mass. In this case, the resultant stiffness coefficient $K_s$ can be determined from the following expression:

$$\frac{1}{K_s} = \frac{1}{K_1} + \frac{1}{K_2}$$

In general, the resultant stiffness coefficient for a sequential system of elastic links reads:

$$\frac{1}{K_s} = \frac{1}{K_1} + \frac{1}{K_2} + \frac{1}{K_3} + \ldots + \frac{1}{K_n} \qquad (2.3)$$

where $i = 1, 2, 3, \ldots, n$, and $n$ is the number of links.

In this text, when physical springs are not present, the mathematical models of motion use springs to represent the stiffness forces. Figure 2.6 shows a combination of two springs with stiffness

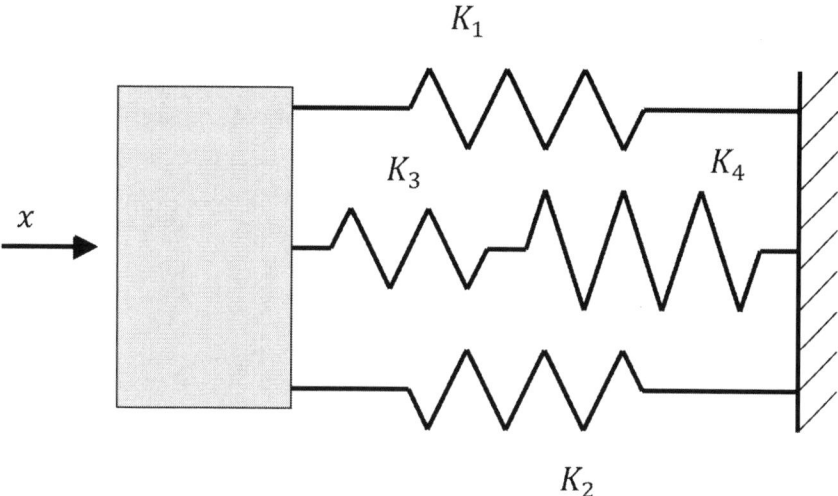

$K_1$

$K_3$          $K_4$

$x$

$K_2$

**Figure 2.6  A combination of springs acting in parallel and in sequential order.**

coefficients $K_1$ and $K_2$ that act in parallel and a pair of springs with stiffness coefficients $K_3$ and $K_4$ that act sequentially. For this case, the resulting stiffness coefficient equals:

$$K = K_1 + K_2 + \frac{K_3 k_4}{K_3 + K_4}$$

For a general case, the resulting stiffness coefficient equals:

$$K = K_p + K_s \qquad (2.4)$$

In the mathematical models presented throughout this book, the springs symbolize the stiffness forces that are exerted during the deformation of either elastic media or elastic links that comprise actual springs. In all cases, we assume that the springs do not possess any mass.

Note that gravitation and magnetism are also associated with forces that depend on the position or displacement.

### 2.1.4  Constant Resisting Forces

In the differential equations of motion, the constant forces may be considered as constant coefficients at time to the zeroth power. The deformation of an elasto-plasic medium in the stage of residual (plastic) deformation is associated with constant resisting forces. Assume that constant resisting forces could exist in some other cases. Therefore, gravity is a constant resistance force in cases of upward motion, but an active force in downward motion.

### 2.1.5  Friction Forces

Dry friction forces have a constant value that equals the product of multiplying a normal force by the friction coefficient. In mechanical engineering, the normal force can be the gravitational force or any other constant force. Friction forces are presented in a separate section because their behavior differs from the behavior of all other constant forces.

The main difference between friction forces and other constant forces is that friction forces always act in the direction opposite to the velocity of motion. The gravity force is directed down regardless of the direction of the velocity. By contrast, the friction force instantaneously changes its direction when the velocity changes its direction.

This feature imposes certain limitations on the applicability of the differential equations of motion that involve friction forces. For example, the differential equation of motion for a vibratory system on a frictional horizontal surface is valid just to the moment when the velocity of the mass becomes equal to zero and immediately starts to move into the opposite direction. The friction force instantaneously changes its direction. However, the equation cannot by itself change the direction of the friction force. This equation becomes invalid for describing the motion of the mass into the opposite direction. A new differential equation of motion is needed to describe the continuation of the motion.

Again, the previous equation cannot account for the instantaneous change in the friction force's direction. If the same vibratory system acts in the vertical direction without the friction force, the system's gravity force would be directed down all the time. The same differential equation would describe the motion of the mass in each direction. See Chapter 3 for an example illustrating this concept.

## 2.2  Analysis of Active Forces

The active forces that are sometimes called the *external forces* or the *applied forces* are usually defined during the formulation of the problem. There is no need to carry out an investigation in order to determine the characteristics of these forces; they are the given forces.

A special case represents the motion of a body as a result of a collision or impact. In such a case, the body obtains a certain initial velocity that can be measured by appropriate devices. Here, the right side of the differential equation of motion (in cases of a collision or impact) equals zero, whereas the initial conditions of motion include the initial velocity obtained from the collision or impact. An appropriate example is presented in Chapter 3.

If active forces cause deceleration of the mechanical system, the forces in the right side of equation (1.6) should have a negative sign.

### 2.2.1  Constant Active Forces

There are no comments specific to the external constant forces. The formulating statements of the engineering problem provide sufficient information about their characteristics.

### 2.2.2  Sinusoidal Active Forces

Forced vibrations are usually caused by sinusoidal active forces. However, vibrations can be developed by constant frictional forces, as in cases of vibrations of a violin string excited by the bow, or vibrations of an air wire due to the wind.

### 2.2.3  Active Forces Depending on Time

Active forces depending on time have a very limited application in movable mechanical systems. They may be caused by using some specific programmable devices.

### 2.2.4  Active Forces Depending on Velocity

Active forces can be functions of the velocity of the motion of some mechanical systems. Consider this example of an active force that decreases due to the increase of the moving mass's velocity. In pneumatically operating systems like jack hammers, soil compactors, and soil penetrating machines, the striker (piston) is accelerated by compressed air. The compressed air is delivered through very small ducts to the chamber behind the striker (the rear chamber), which has a relatively big cross-sectional area. When the striker's motion starts, the air pressure in the rear chamber is close to the pressure in the ducts. However, when the velocity of the striker rapidly increases, the air pressure in the rear chamber continually decreases. This decrease happens because the relatively small ducts restrict the air flow into the rear chamber. In turn, the supply of compressed air cannot catch up with the rapidly increasing volume of this chamber, causing in drop of the pressure in the rear chamber and, consequently, the decrease of the force applied to the striker. Thus, the applied pressure force depends on the velocity of the striker and should be considered as an active or external damping force that causes motion. There are other examples where damping forces represent active forces, including the example of acceleration of an automobile's rear wheel drive, found in Chapter 4.

When solving the differential equation of motion, we combine the resisting damping force and the active damping force into one damping force. An example is presented in Chapter 4.

### 2.2.5  Active Forces Depending on Displacement

In some situations, the active forces are functions of the mass's displacement. Let's consider a certain interval of motion of an internal combustion engine's piston.

At the instant the fuel mixture ignites, when the piston is in its upper point, the combustion chamber has its minimum volume. The pressure at this moment reaches its highest value; consequently, the largest active force is applied to the piston. The piston's displacement leads to the volume increase of the chamber, although the amount of the compressed gases remains the same. In this case, the volume increase causes the pressure to decrease, which, in turn, results in a decrease of the active force applied to the piston. A similar process occurs in the barrel of fire arms, where the displacement of the projectile causes the drop of pressure in the barrel, with a consequent decrease of the active force applied to the projectile.

In solving the differential equation of motion, we combine the resisting stiffness force with the active stiffness force into one combined stiffness force. An example in Chapter 4 illustrates these considerations.

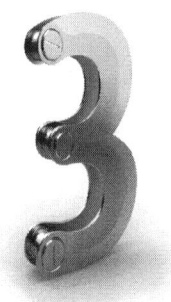

# SOLVING DIFFERENTIAL EQUATIONS
# OF MOTION USING
# LAPLACE TRANSFORMS

The Laplace Transform provides a straightforward, universal methodology for solving linear differential equations of motion as well as systems of these equations. This methodology is based on converting a differential equation of motion from the time domain into a corresponding equation in the Laplace domain. After the conversion, the new Laplace domain equation is made up entirely of conventional algebraic expressions.

Some of these expressions in the left side of this equation have the Laplace domain displacement function as multipliers. This displacement function represents the solution of the Laplace domain equation; the function should be factored out using regular algebraic procedures. By completing these procedures, the Laplace domain equation will have in its left side the displacement function in the Laplace domain while in its right side a summation/subtraction of proper algebraic fractions. These fractions represent functions of the

Laplace domain independent variable. The final step of this methodology is to invert both parts of this equation into the time domain. The result of this inversion represents the solution of the differential equation of motion in the time domain.

The processes of conversion and inversion are based on identifying expressions and their counterparts in both domains; these pairs are called Laplace pairs. They are tabulated in many related sources. Pairs are identified by comparing their structures; no specific mathematical actions are necessary. Note that $x(t)$ is the displacement as a function of the running time in the time domain whereas $x(l)$ is the corresponding displacement function in the Laplace domain, with $l$ is an independent variable complex value.

You do not need to memorize the detailed basics of the Laplace Transform in order to use its methodology. Even though the fundamentals of the Laplace domain are presented in numerous publications, there is no need to get familiar with these fundamentals in order to comprehend the contents of this book and widely use the Laplace Transform to solve linear differential equations of motion.

The Laplace Transform methodology of solving differential equations has three steps:

1. Convert the differential equation from the time domain into the Laplace domain.
2. Solve the converted equation for the Laplace domain displacement function using conventional algebraic operations.
3. Invert the solution from the Laplace domain into the time domain form.

Step-by-step procedures of this method for solving differential equations of motion are presented bellow.

## 3.1  Laplace Transform Pairs for Differential Equations of Motion

The conversion and inversion procedures required for solving differential equations are based on identifying the equivalent Laplace

Transform pairs from the corresponding tables. Numerous catalogs are available with tables of many hundreds of Laplace Transform pairs. Here are a few examples of Laplace Transform pairs:

The time domain displacement function $x(t)$ is equivalent to the Laplace domain displacement function $x(l)$. Thus, the corresponding Laplace pair is:

$$x(t) \rightarrow x(l)$$

The time domain independent variable, the running time $t$, is equivalent to the independent variable $\frac{1}{l}$ in the Laplace domain. The corresponding Laplace pair is:

$$t \rightarrow \frac{1}{l}$$

Actually, the unit of $l$ is $\frac{1}{s}$

A constant value or a constant coefficient in the time domain is the same in the Laplace domain. In this case, the Laplace pair is:

$$C \rightarrow C$$

In order to simplify the process of solving differential equations of motion by using Laplace Transforms, I have compiled a table of the most applicable Laplace Transform pairs for the majority of real-life mechanical engineering problems (see Table 3.1). The left column of this table lists the time domain functions while the right column lists the Laplace domain functions. A Laplace Transform pair represents an expression in the time domain that is equivalent to the expression in the Laplace domain. These expressions can be converted or inverted into each other. In the majority of cases when the expressions are not too long, the Laplace Transform pairs are placed in the same line. The lines and, consequently, the pairs in the table are numbered and each pair has the same ordinal number.

| Table 3.1 Laplace Transform Pairs | | | |
|---|---|---|---|
| **Time domain functions** | | **Laplace domain functions** | |
| 1. | $x(t)$ or $x$, or $\theta(t)$, or $\theta$ | 1. | $x(l)$ or $\theta(l)$ |
| 2. | constant | 2. | constant |
| 3. | t | 3. | $\dfrac{1}{l}$ |
| 4. | $\dfrac{t^n}{n!}$ | 4. | $\dfrac{1}{l^n}$ <br><br> for pair #4, where $n$ is a positive integer |
| 5. | $\dfrac{e^{nt}-1}{n}$ | 5. | $\dfrac{1}{l-n}$ |
| 6. | $e^{-nt}$ | 6. | $\dfrac{l}{l+n}$ |
| 7. | $\dfrac{1}{\omega^2}(1-\cos\omega t)$ | 7. | $\dfrac{1}{l^2+\omega^2}$ |
| 8. | $\sin\omega t$ | 8. | $\dfrac{\omega l}{l^2+\omega^2}$ |
| 9. | $\cos\omega t$ | 9. | $\dfrac{l^2}{l^2+\omega^2}$ |
| 10. | $te^{-nt}$ | 10. | $\dfrac{l}{(l+n)^2}$ |
| 11. | $e^{-nt}\sin\omega t$ | 11. | $\dfrac{\omega l}{(l+n)^2+\omega^2}$ |
| 11a. | $e^{-nt}\sinh\omega t$ | 11a. | $\dfrac{\omega l}{(l+n)^2-\omega^2}$ |
| 12. | $e^{-nt}\cos\omega t$ | 12. | $\dfrac{l(l+n)}{(l+n)^2+\omega^2}$ |
| 12a. | $e^{-nt}\cosh\omega t$ | 12a. | $\dfrac{l(lp+n)}{(l+n)^2-\omega^2}$ |

| | | | |
|---|---|---|---|
| 13. | $\dfrac{1}{\omega^2+n^2}[1-e^{-nt}$ $\times(\cos\omega t+\dfrac{n}{\omega}\sin\omega t)]$ | 13. | $\dfrac{1}{(l+n)^2+\omega^2}$ |
| 13a. | $\dfrac{1}{\omega^2+n^2}[1-e^{-nt}$ $\times(\cosh\omega t+\dfrac{n}{\omega}\sinh\omega t)]$ | 13a. | $\dfrac{1}{(l+n)^2-\omega^2}$ |
| 14. | $\dfrac{1}{\omega}e^{-nt}\sin\omega t$ | 14. | $\dfrac{l}{(l+n)^2+\omega^2}$ |
| 14a. | $\dfrac{1}{\omega}e^{-nt}\sinh\omega t$ | 14a. | $\dfrac{l}{(l+n)^2-\omega^2}$ |
| 15. | $(\cos\omega t-\dfrac{n}{\omega}\sin\omega t)e^{-nt}$ | 15. | $\dfrac{l^2}{(l+n)^2+\omega^2}$ |
| 16. | $\dfrac{dx}{dt}$ or $\dfrac{d\theta}{dt}$ | 16. | $lx(l)-ls_0$ or $l\theta(l)-l\theta_0$ for pair #16, where $x_0$ and $\theta_0$ are respectively the initial displacement or initial angular displacement |
| 17. | $\dfrac{d^2x}{dt^2}$ or $\dfrac{d^2\theta}{dt^2}$ | 17. | $l^2x(l)-lv_0-l^2s_0$ or $l^2\theta(l)-l\Omega_0-l^2\theta_0$ for pair #17, where $x_0$ and $\theta_0$ are respectively the initial displacement or initial angular displacement, and $v_0$ and $\Omega_0$ are respectively the initial velocity or initial angular velocity |
| 18. | $\dfrac{\omega\sin\varphi t-\varphi\sin\omega t}{\varphi\omega(\omega^2-\varphi^2)}$ | 18. | $\dfrac{l}{(l^2+m^2)(l^2+\varphi^2)}$ |
| 19. | $\dfrac{1}{\omega^2}(t-\dfrac{1}{\omega}\sin\omega t)$ | 19. | $\dfrac{1}{l(l^2+\omega^2)}$ |
| 20. | $\dfrac{1}{\omega^4}(\cos\omega t-1)+\dfrac{1}{2\omega^2}t^2$ | 20. | $\dfrac{1}{l^2(l^2+\omega^2)}$ |

| | | | |
|---|---|---|---|
| 21. | $te^{-nt}$ | 21. | $\dfrac{l}{(l+n)^2}$ |
| 22. | $(1-nt)e^{-nt}$ | 22. | $\dfrac{l^2}{(l+n)^2}$ |
| 23. | $\dfrac{1}{\omega^4}(1-\cos\omega t)-\dfrac{1}{2\omega^3}t\sin\omega t$ | 23. | $\dfrac{1}{(l^2+\omega^2)^2}$ |
| 24. | $\dfrac{1}{2\omega^3}\sin\omega t-\dfrac{t}{2\omega^2}\cos\omega t$ | 24. | $\dfrac{l}{(l^2+\omega^2)^2}$ |
| 25. | $\dfrac{1}{\sqrt{(\delta^2-\varepsilon^2)^2+4n^2\varepsilon^2}}$ $\times[\dfrac{1}{\varepsilon}\sin(\varepsilon t-\gamma)$ $+\dfrac{1}{\omega}e^{-nt}\sin(\omega t-\mu)]$ $\gamma=tan^{-1}\left(\dfrac{2n\varepsilon}{\delta^2-\varepsilon^2}\right)$ $\mu=tan^{-1}\left(\dfrac{-2n\omega}{\delta^2-\omega^2+\varepsilon^2}\right)$ $\delta^2=n^2+\omega^2$ | 25. | $\dfrac{1}{(l^2+\varepsilon^2)[(l+n)^2+\omega^2]}$ |
| 26. | $\dfrac{1}{n^2+\omega^2}[t-\dfrac{2n}{n^2+\omega^2}$ $+\dfrac{1}{\omega}e^{-nt}\sin(\omega t-\varphi)]$ $\varphi=2\tan^{-1}\left(-\dfrac{\omega}{n}\right)$ | 26. | $\dfrac{1}{l[(l+n)^2+\omega^2]}$ |
| 27. | $\dfrac{1}{\sqrt{(\varepsilon^2-\omega_1^2)^2+4n^2\omega_1^2}}$ $\times[\dfrac{1}{\omega_1}\sin(\omega_1 t-\delta)$ $+\dfrac{1}{\omega}e^{-nt}\sin(\omega t-\sigma)]$ $\delta=\tan^{-1}(\dfrac{2n\omega_1}{\varepsilon^2-\omega_1^2})$ $\sigma=\tan^{-1}(\dfrac{-2n\omega}{n^2-\omega^2+\omega_1^2})$ $\varepsilon^2=n^2+\omega^2$ | 27. | $\dfrac{l}{\left(l^2+\omega_1^2\right)[(l+n)^2+\omega^2]}$ |

Note: Hyperbolic functions
1. Hyperbolic sine: $sinh t=\frac{e^t-e^{-t}}{2}$  (3a)
2. Hyperbolic cosine: $cosh t=\frac{e^t+e^{-t}}{2}$  (3b)

As described above, the method of using the Laplace Transform to solve linear differential equations of motion has three steps.

1. In the first step, we convert all the terms of the differential equation of motion from the time domain into the Laplace domain. This conversion is based on identifying the appropriate Laplace Transform pairs. All possible time domain terms can be found in the left column of Table 3.1. Converting the equation is a straightforward process that involves trivial operations, such as finding in the left column the expressions that are similar to the expressions in the differential equation of motion. The counterpart expressions from the right column indicate the Laplace domain. The differential equation of motion converted into the Laplace domain represents an algebraic expression.
2. In the second step, we solve the converted expression for the Laplace domain displacement function. This step is based on conventional algebraic procedures that lead to an equation with the Laplace domain displacement function on the left side and a single algebraic proper fraction (or a sum of these fractions) on the right.
3. In the third step, we invert both sides of the solution from the Laplace domain into the time domain, using Table 3.1 in reverse order. The Laplace domain displacement from the left side of the solution simply inverts into the displacement function in the time domain. In most cases, the fractions in the right side have equivalent representations in the right column of Table 3.1; they are inverted by using corresponding pairs in the time domain.

The result from this third step represents the solution of the differential equation of motion. However, occasionally one or more of the fractions in the right side of the Laplace domain equation are not found in the right column of Table 3.1 — nor in other appropriate sources. The reason is that these missing fractions can be resolved into simpler fractions found in Table 3.1. Thus, when the fractions are missing in Table 3.1, apply some additional algebraic procedures to resolve them into an equivalent sum of fractions that are tabulated. These procedures are associated with the process of decomposition of fractions and are demonstrated below.

The use of the expressions in Table 3.1 will become apparent during the process of solving differential equations of motion. Numerous examples are presented in the following sections. However, before getting familiar with the methodology of solving differential equations of motion, it is necessary to consider the procedures of decomposing fractures that are not represented in Table 3.1.

## 3.2  Decomposition of Proper Rational Fractions

We've described above the three steps we take to solve linear differential equations of motion with the help of Laplace Transforms. The third step, inverting the fractions into the time domain, is based on identifying the appropriate pairs in Table 3.1. It may happen that not all of these proper rational fractions have appropriate representations in the table.

In such cases, the fractions should undergo the process of decomposition (expansion) that transforms them into a sum of simpler fractions which are included in Table 3.1. The same procedure of decomposition is used when solving integrals. Numerous sources outside this text present the principles and procedures of decomposition. However, it is desirable to have in one place all the information needed for solving differential equations of motion. Thus, some of the basic algebraic procedures related to the decomposition of proper fractions are presented below.

Let's consider a proper rational fraction $\frac{f(l)}{F(l)}$ that is not found in Table 3.1. In the most general case, this fraction may have the following structure:

$$\frac{f(l)}{F(l)} = \frac{f(l)}{l(l+a)(l^2+b^2)\left[(l+c)^2+d^2\right]} \qquad (3.1)$$

where $a$, $b$, $c$, and $d$ are constant values, and $l$ is the independent variable in the Laplace domain. Note that the independent variable could be the time $t$ or other parameters. The fraction in the right side of equation (3.1) is not found in Table 3.1. Consequently, it cannot be converted directly into the time domain. The purpose of decomposition is to resolve this fraction into a sum of simpler fractions that are each found in Table 3.1.

The decomposition of this fraction leads to the following sum of proper fractions:

$$\frac{f(l)}{F(l)} = \frac{A}{l} + \frac{B}{l+a} + \frac{C+Dl}{l^2+b^2} + \frac{E+Hl}{(l+c)^2+d^2} \qquad (3.2)$$

where $A$, $B$, $C$, $D$, $E$, and $H$ are constant coefficients that should be determined. The decomposition is based on structuring a sum of fractions so that their denominators respectively represent the single multipliers of the denominator of the initial fraction. The values of the constant coefficients can be calculated in several ways. However, the most universal way is the method of undetermined coefficients. The examples below include this method of determining the coefficients.

## 3.3  Examples of Decomposition of Fractions

The examples below show the step-by-step procedures of decomposing proper rational fractions into corresponding sums of simpler fractions that usually are found in Table 3.1. They also show ways to structure the numerators and denominators of the simpler fractions in accordance with the initial fractions. These examples also demonstrate how to calculate constant coefficients.

## Example 1

$$\text{Given: } \frac{f(l)}{F(l)} = \frac{3l+2n}{l(l-n)}$$

where $n$ represents a constant value (both in this and the following examples).

The fraction in the right side of this equation is not found in Table 3.1.

## Decomposition

Based on equation (3.2), we have:

$$\frac{3l+2n}{l(l-n)} = \frac{A}{l} + \frac{B}{l-n} \qquad (3.3)$$

Bringing the right side of equation (3.3) to the lowest common denominator, we write:

$$\frac{3l+2n}{l(l-n)} = \frac{A(l-n)+Bl}{l(l-n)} \qquad (3.4)$$

Omitting the denominators in both sides of equation (3.4), we have:

$$3l+2n = A(l-n)+Bl = Al - An + Bl$$

Applying the appropriate algebra to this equation, we finally may write:

$$3l+2n = l(A+B) - An \qquad (3.5)$$

Using the method of undetermined coefficients, we compose a system of simultaneous equations that allows us to calculate the unknown coefficients. In this case, we have two unknown coefficients $A$ and $B$; consequently, we need a system of two equations. This system can be obtained by equating the coefficients at the same powers of $l$ in both sides of equation (3.5). The coefficient of $l$ in the left side of equation (3.5) is 3, whereas in the right side it is $A+B$. Thus, the first equation reads:

$$3 = A+B \qquad (3.6)$$

For the constant values, the parameter $l$ is to the zeroth power. Thus, we just equate the constant values of both parts of the equation. The constant value in the left side of equation (3.5) is $2n$ whereas in the right side it is $-An$. So, we have the second equation:

$$2n = -An \qquad (3.7)$$

Equations (3.6) and (3.7) are examples of using the method of undetermined coefficients. Solving these two equations simultaneously, we find that $A = -2$ and $B = 5$. Substituting the values of these coefficients into the equation (3.3), we have:

$$\frac{3\,l + 2n}{l(l - n)} = -\frac{2}{l} + \frac{5}{l - n} \qquad (3.8)$$

The fractions in the right side of equation (3.8) are represented in the right column of Table 3.1 by expressions 3 and 5 respectively.

**Example 2**

Given: $\dfrac{f(l)}{F(l)} = \dfrac{l + 2n}{l(l + n)^2}.$

The fraction in the right side of this equation is not represented in Table 3.1.

**Decomposition**

Referring to equation (3.2), we write:

$$\frac{l + 2n}{l(l + n)^2} = \frac{A}{l} + \frac{B}{l + n} + \frac{Cl}{(l + n)^2} \qquad (3.9)$$

If the denominator comprises the parameter $l$ to the second power, the numerator should have this parameter to the first power.

Omitting the description of the trivial algebraic operations in equation (3.9), we have:

$$l + 2n = l^2\left(A + B + C\right) + l(2An + Bn) + An^2 \qquad (3.10)$$

Applying the method of undetermined coefficients to equation (3.10), we note that the left side does not have a term with $l^2$, which means that the coefficient for $l^2$ is zero. Therefore, we obtain:

$$0 = A + B + C$$

$$1 = 2An + Bn$$

$$2n = An^2$$

Solving this system of three equations simultaneously, we determine:

$$A == \frac{2}{n}, \quad B = -\frac{3}{n}, \quad C = 1,$$

Substituting the values of these coefficients into equation (3.9) we find:

$$\frac{l+2n}{l(l+n)^2} = \frac{2}{nl} - \frac{3}{n(l+n)} + \frac{l}{(l+n)^2} \tag{3.11}$$

Referring to the right column in Table 3.1, we find that the first fraction on the right side of equation (3.11) is represented by expression 3 and the second fraction is represented by expression 5, while keeping in mind that

$$l - (-n) = l + n$$

and finally the third fraction is represented by the expression 10.

**Example 3**

Given: $\dfrac{f(l)}{F(l)} = \dfrac{l+1}{(l-1)[(l+2)^2 + 3^2]}$

The fraction from the right side of this equation is not represented in Table 3.1.

**Decomposition**

$$\frac{l+1}{(l-1)[(l+2)^2 + 3^2]} = \frac{A}{l-1} + \frac{Bl+C}{(l+2)^2 + 3^2} \tag{3.12}$$

After completing the appropriate algebraic operations, we obtain:

$$l+1 = l^2(A+B) + l(4A - B + C) + 13A - C \tag{3.13}$$

Equation (3.13) yields the following system:

$$0 = A + B$$

$$1 = 4A - B + C$$

$$1 = 13A - C$$

Solving this system of three equations simultaneously, we calculate the values of the coefficients:

$$A = \frac{1}{9}, \quad B = -\frac{1}{9}, \quad \text{and} \quad C = \frac{4}{9}$$

Substituting these values into the right side of equation (3.12), we have:

$$\frac{l+1}{(l-1)[(l+2)^2 + 3^2]} = \frac{1}{9(l-1)} - \frac{l-4}{9[(l+2)^2 + 3^2]} \tag{3.14}$$

The first fraction in the right side of equation (3.14) is represented in Table 3.1 by expression 5 in the right column., However, the second fraction is not represented in Table 3.1. In such cases, we may rewrite equation (3.14) in the following way:

$$\frac{l+1}{(l-1)[(l+2)^2 + 3^2]} = \frac{1}{9(l-1)} - \frac{l}{9[(l+2)^2 + 3^2]}$$
$$+ \frac{4}{9[(l+2)^2 + 3^2]} \tag{3.15}$$

Referring to Table 3.1, we can see the second fraction in the right side of the equation is represented by expression 14, whereas the third fraction is represented by expression 13.

More examples of decomposing fractions that cannot be found in tables of Laplace Transform pairs are presented below while solving the differential equations of motion.

### 3.4  Examples of Solving Differential Equations of Motion

In order to demonstrate the methodology of solving the differential equations of motion by using Laplace Transform, we begin with an example of the simplest equation. The complexity of the equations is gradually increased in the following examples.

### 3.4.1  Motion by Inertia with no Resistance

In this hypothetical case, a body having a mass $m$ moves horizontally on a frictionless surface with no resistance. No active or external forces are applied to it. However, it possesses kinetic energy — in other words, the body has a certain initial velocity.

Referring to equation (1.1), we compose the following differential equation of motion of the body for this case:

$$\mathrm{m}\frac{d^2x}{dt^2} = 0 \qquad\qquad (3.16)$$

the initial conditions of motion are:

$$\text{for } t = 0 \quad x = s_0, \quad \frac{dx}{dt} = v_0 \qquad\qquad (3.17)$$

Dividing both sides of equation (3.16) by $m$, we obtain:

$$\frac{d^2x}{dt^2} = 0 \qquad\qquad (3.18)$$

Now we apply the first step: converting equation (3.18) into the Laplace domain. Referring to the left column of Table 3.1, we find that according to expression 17 the second derivative is the same as the left side of equation (3.18). Consequently, expression 17 in the right column of Table 3.1 should be used to convert the second derivative. The structure of this expression requires the inclusion of the corresponding initial conditions of motion — namely, the initial velocity and the initial displacement.

As will be seen later, expression 16 in the right column, which is used to convert the first derivative, requires the inclusion of the initial displacement. Thus, expressions 16 and 17 are the only places to include the initial conditions of motion, while solving the differential equation of motion using the Laplace Transform.

Because all differential equations of motion include the second derivative, any conversion of these equations should start with the 17th pair of expressions from Table 3.1.

The right side of equation (3.18) is zero, and its conversion is also zero. Thus, we may write:

$$l^2 x(l) - l v_0 - l^2 s_0 = 0 \tag{3.19}$$

During the second step, we solve equation (3.19) for the displacement in the Laplace domain $x(l)$. Dividing both parts of the equation (3.19) by $l^2$ we obtain:

$$x(l) - \frac{v_0}{l} - s_0 = 0 \tag{3.20}$$

Applying algebraic operations, we have:

$$x(l) = \frac{v_0}{l} + s_0 \tag{3.21}$$

Applying the third step, we invert equation (3.21) into the time domain using the first pair of expressions in Table 3.1. The numerator $v_0$ of the right-side of the fraction represents a constant value and remains the same after inversion. Therefore, a constant value should be inverted by using the second pair of expressions. Keep in mind that the constant coefficients in both domains are the same. The fraction $\frac{1}{l}$ should be inverted by using the third pair of expressions. The second term in the right side of equation (3.21) is also a constant value and is also inverted according to the second pair of expressions. Finally, we obtain:

$$x = v_0 t + s_0 \tag{3.22}$$

Thus, equation (3.22) represents the solution of differential equation (3.18) with its initial conditions (3.17). Equation (3.22) is very well known from basic physics; it describes the uniform motion of a rigid body.

### 3.4.2    Motion by Inertia with Resistance of Friction

In contrast to the previous case, a frictional resisting force is applied to a body that is also moving horizontally and also possesses a certain initial velocity. No active forces are applied to the body. Referring to equation (1.1), we have the following differential equation of motion:

$$m\frac{d^2x}{dt^2} + F = 0 \tag{3.23}$$

The initial conditions are the same as in the previous example, according to expression (3.17). Dividing both sides of equation (3.23) by $m$ we write:

$$\frac{d^2x}{dt^2} + f = 0 \tag{3.24}$$

where

$$f = \frac{F}{m} \tag{3.25}$$

During the first step, and using pairs 2 and 17 from Table 3.1, we convert equation (3.24) with its initial conditions (3.17) into the Laplace domain:

$$l^2x(l) - lv_0 - l^2s_0 + f = 0 \tag{3.26}$$

The second step is associated with the solution of equation (3.26) for displacement in the Laplace domain $x(l)$:

$$x(l) = s_0 + \frac{v_0}{l} - \frac{f}{l^2} \tag{3.27}$$

Completing the third step, we invert equation (3.27) into the time domain by using respectively pairs 1, 2, 3, and 4 from Table 3.1. Thus, we obtain:

$$x = s_0 + v_0 t - \frac{ft^2}{2} \tag{3.28}$$

Equation (3.28) represents the solution of the differential equation of motion (3.24) with the initial conditions (3.17). This equation of motion actually describes the process of deceleration of the body.

### 3.4.3   Motion by Inertia with Damping Resistance

Figure 3.1 represents a system where the mass $m$ moves by inertia horizontally on a frictionless surface experiencing damping resistance. The mass is attached to a dashpot, which has a damping coefficient $C$ and exerts a damping force that slows down the motion of the mass. As in the two previous cases, the mass of the system possesses a certain initial velocity and no active forces are applied to the system.

Based on the model in Figure 3.1 and referring to equation (1.1), we write the following differential equation of motion for this case:

$$m\frac{d^2 x}{dt^2} + C\frac{dx}{dt} = 0 \tag{3.29}$$

where $C$ is the damping coefficient.

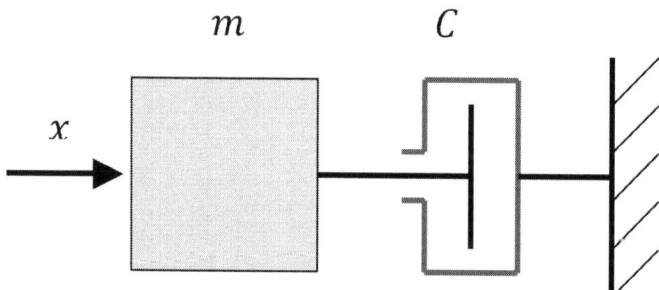

**Figure 3.1 The model of a damped motion.**

As with previous cases, the initial conditions of motion are presented by expression (3.17). Dividing both parts of equation (3.29) by $m$, we obtain:

$$\frac{d^2x}{dt^2} + 2n\frac{dx}{dt} = 0 \qquad (3.30)$$

where $n$ is the damping factor, and as will be seen later, it is convenient to use the following notation:

$$2n = \frac{C}{m} \qquad (3.31)$$

Based on the pairs of expressions 2, 16, and 17 from Table 3.1, we convert equation (3.30) with its initial conditions (3.17) into the Laplace domain:

$$l^2x(l) - lv_0 - l^2s_0 + 2nlx(l) - 2nls_0 = 0$$

Solving this equation for $x(l)$, we have:

$$x(l) = \frac{v_0 + 2ns_0 + ls_0}{l + 2n} \qquad (3.32)$$

The right side of equation (3.32) is not represented in Table 3.1; however, this equation can be expressed in the following way:

$$x(l) = \frac{v_0 + 2ns_0}{l - (-2n)} + \frac{ls_0}{l + 2n} \qquad (3.33)$$

The numerator of the first fraction in equation (3.33) represents a constant value; the structures of the fractions are identical to ones available in Table 3.1

In order to invert equation (3.33) into the time domain, we use the pairs of expressions 1, 5, and 6 from Table 3.1. The constant values are inverted by using pair 2 — we will not continue to emphasize this for the successive examples. Thus, we write:

$$x = (v_0 + 2ns_0)(\frac{e^{-2nt} - 1}{-2n}) + s_0e^{-2nt} \qquad (3.34)$$

Analyzing equation (3.34) further, we obtain:

$$x = \frac{1}{2n}\left(2ns_0 + v_0 - v_0 e^{-2nt}\right) \qquad (3.35)$$

Equation (3.35) is the solution of the differential equation of motion (3.29) with its initial conditions (3.17); it represents the law of motion of the considered system. The solutions of the previous and successive differential equations of motion for the mechanical systems are also the laws of motion for these systems.

### 3.4.4 Free Vibrations

The structure of a simple vibratory system represents a mass attached to a spring that is secured to a non-movable support. Let's consider the motion of the mass on a frictionless horizontal surface. The mass is not subjected to the action of external forces. In this case, if the mass possesses a certain velocity, its motion could be caused by kinetic energy, or by the potential energy of deformation of the spring, or by the combined action of both energy factors. The mass will perform vibratory motion that is called *free vibrations*. If both parameters of the initial conditions of motion (the displacement and the velocity) equal zero, and no external force is applied, there will be no motion.

The initial conditions of motion determine the cause of motion. When the initial displacement has a certain value while the initial velocity equals zero, the spring is deformed proportionally to this displacement. The motion then occurs due to the potential energy of the deformed spring. However, when the initial displacement equals zero while the initial velocity has a certain value, the motion occurs due to the kinetic energy that the mass possesses. In other cases, when the initial parameters of motion differ from zero, the motion is caused by both the kinetic and potential energy.

The model of free vibrations of a mass on a frictionless horizontal surface is shown in Figure 3.2. When the body is in position 1 at the origin O of the axis $x$, the spring is not deformed. In position 2, however, the mass has an initial displacement $s_0$, and the initial velocity of the mass equals $v_0$. The description of the free vibratory motion of the mass begins from position 2. According to these

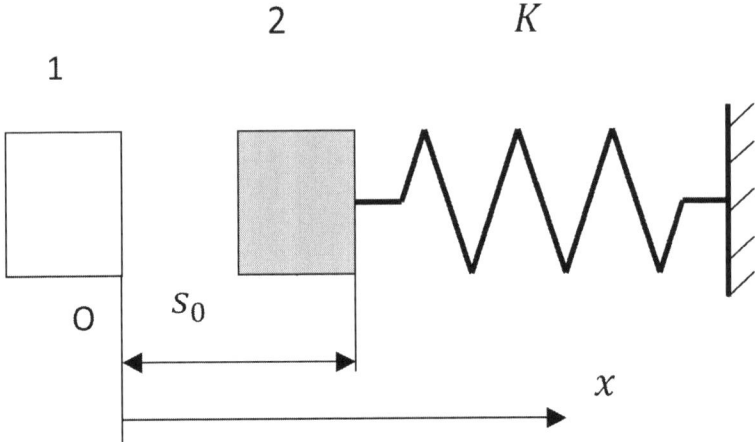

**Figure 3.2  The model of free undamped vibrations.**

considerations, and referring to both equation (1.1) and Figure 3.2, we can compose the following differential equation of motion:

$$m\frac{d^2x}{dt^2} + Kx = 0 \tag{3.36}$$

Again, the initial conditions of motion are presented by expression (3.17). Dividing equation (3.36) by $m$, we have:

$$\frac{d^2x}{dt^2} + \omega_0^2 x = 0 \tag{3.37}$$

In equation (3.37), $\omega_0$ is the natural frequency of the system and it equals:

$$\omega_0^2 = \frac{K}{m} \tag{3.38}$$

Using pairs 1, 2, and 17 from Table 3.1, we convert equation (3.37) with the initial conditions (3.17) into the Laplace domain:

$$l^2 \mathrm{x}(l) - lv_0 - l^2 s_0 + \omega_0^2 x(l) = 0 \tag{3.39}$$

Solving equation (3.39) for $x(l)$, we write:

$$x(l) = \frac{lv_0}{l^2 + \omega_0^2} + \frac{l^2 s_0}{l^2 + \omega_0^2} \tag{3.40}$$

Applying pairs 1, 8, and 9 from Table 3.1 to equation (3.40), we invert it into the time domain:

$$x = \frac{1}{\omega_0}(v_0 \sin\omega_0 t + s_0\omega_0 \cos\omega_0 t) \tag{3.41}$$

Equation (3.41) describes the free vibrations. Applying to this equation some conventional transformations, we can present it in a form that is more suitable for analysis. Multiplying and dividing the right side of equation (3.41) by $\sqrt{v_0^2 + s_0^2\omega_0^2}$, we may write:

$$x = \frac{\sqrt{v_0^2 + s_0^2\omega_0^2}}{\omega_0}\left(\frac{v_0}{\sqrt{v_0^2 + s_0^2\omega_0^2}}\sin\omega_0 t + \frac{s_0\omega_0}{\sqrt{v_0^2 + s_0^2\omega_0^2}}\cos\omega_0 t\right) \tag{3.42}$$

The radical includes the sum of the squares of the coefficients from the trigonometric functions of equation (3.41). Consider the following identity:

$$\frac{v_0^2}{v_0^2 + s_0^2\omega_0^2} + \frac{s_0^2\omega_0^2}{v_0^2 + s_0^2\omega_0^2} = (\sin\alpha)^2 + (\cos\alpha)^2 = 1$$

Based on this expression we denote:

$$\frac{v_0}{\sqrt{v_0^2 + s_0^2\omega_0^2}} = \sin\alpha \tag{3.43}$$

$$\frac{s_0\omega_0}{\sqrt{v_0^2 + s_0^2\omega_0^2}} = \cos\alpha \tag{3.44}$$

Substituting equations (3.43) and (3.44) into equation (3.42), we obtain:

$$x = \frac{\sqrt{v_0^2 + s_0^2\omega_0^2}}{\omega_0}(\sin\omega_0 t \sin\alpha + \cos\omega_0 t \cos\alpha) \qquad \textbf{(3.45)}$$

Finally we have:

$$x = \frac{\sqrt{v_0^2 + s_0^2\omega_0^2}}{\omega_0}\cos(\omega_0 t - \alpha) \qquad \textbf{(3.46)}$$

In comparison with equation (3.45), the structure of equation (3.46) is more suitable for the appropriate analysis. For instance, let us determine the time $t^*$ at which the velocity of the mass becomes equal to zero and, consequently, at which the acceleration gets its maximum value. Taking the first derivative from equation (3.46), we write:

$$\frac{dx}{dt} = -\sqrt{v_0^2 + s_0^2\omega_0^2}\ \sin(\omega_0 t - \alpha) \qquad \textbf{(3.47)}$$

Equating the left side of equation (3.47) to zero we have:

$$\sin(\omega_0 t^* - \alpha) = 0 \qquad \textbf{(3.48)}$$

Based on equation (3.48), we may write:

$$\omega_0 t^* - \alpha = i\pi, \quad i = 0, 1, 2, 3,\ldots$$

From this expression, we calculate:

$$t^* = \frac{i\pi + \alpha}{\omega_0} \qquad \textbf{(3.49)}$$

Differentiating equation (3.47), we obtain the expression for the acceleration:

$$\frac{d^2x}{dt^2} = -\omega_0 \sqrt{v_0^2 + s_0^2\omega_0^2} \, \cos(\omega_0 t - \alpha)$$

Substituting into this equation the time according to equation (3.49), we determine the maximum acceleration $a_{max}$ at $i = 1$:

$$a_{max} = \omega_0 \sqrt{v_0^2 + s_0^2\omega_0^2}$$

Using equation (3.42) would require many more transitional steps in order to obtain the same results. Where it is applicable, it would be justifiable to transform the equations of motion into the format similar to equation (3.46).

### 3.4.5    Motion Caused by Impact

We consider an impact, a blow, a strike, or a collision to be a physical phenomenon characterized by applying a force to a rigid body; that force tends toward infinity during an infinitesimal amount of time. The impact represents the change in the momentum of the body. It equals the product of multiplying the force by the time. In mathematics, the product of multiplying an infinite value by an infinitesimal one results in a finite value. Thus, the impact is a finite value.

There are no ways to measure the magnitude of an infinite force or an infinitesimal time. However, the change of the body's momentum as a result of an impact can be determined experimentally. A momentum equals the product of multiplying the mass of the body by its velocity. The mass cannot be changed due to an impact. Thus, the impact results in the change to the body's velocity.

Measuring the value of the mass is a trivial procedure. Fortunately, contemporary measuring instruments let us determine the instantaneous change of the body's velocity due to an impact. Estimating the magnitude of an impact is based on determining the change of the mass's velocity due to the impact. These considerations let us set the right side of the differential equation of motion equal to zero while accounting for the result of the impact by an appropriate change of the mass's initial velocity.

Let's consider an example of motion of a rigid body due to an impact. Figure 3.3 shows a model of a system that includes a mass, an elastic link, and a dashpot. This model represents a case of free damped vibrations. As a result of an impact, the mass starts to move horizontally on a frictionless surface.

The mass obtains an initial velocity $v_0$. Referring to both equation (1.1) and Figure 3.3, we assemble the following differential equation of motion:

$$m\frac{d^2x}{dt^2} + C\frac{dx}{dt} + Kx = 0 \qquad (3.50)$$

The initial conditions of motion are:

$$\text{for} \quad t = 0 \quad x = 0, \quad \frac{dx}{dt} = v_0 \qquad (3.51)$$

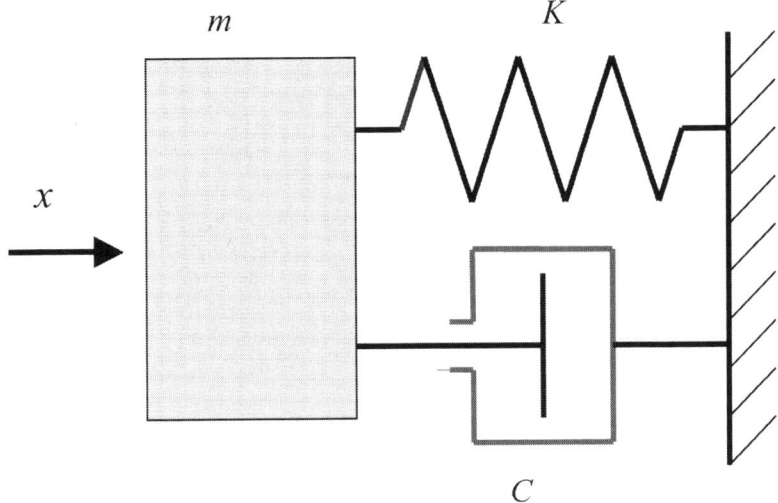

**Figure 3.3 A model of a damped vibratory system.**

Dividing equation (3.50) by $m$ and using equations (3.31) and (3.38), we may write:

$$\frac{d^2x}{dt^2} + 2n\frac{dx}{dt} + \omega_0^2 x = 0 \qquad (3.52)$$

Accounting the initial conditions (3.51) and using pairs 1, 2, 16, and 17 from Table 3.1, we convert equation (3.52) into the Laplace domain:

$$l^2 x(l) - lv_0 + 2nlx(l) + \omega_0^2 x(l) = 0 \qquad (3.53)$$

Solving equation (3.53) for $x(l)$, we have:

$$x(l) = \frac{lv_0}{l^2 + 2nl + \omega_0^2} \qquad (3.54)$$

In order to invert equation (3.54) into the time domain, we should restructure the denominator of the fraction in the right side of the equation in the following way:

$$l^2 + 2nl + \omega_0^2 + n^2 - n^2 = l^2 + 2nl + n^2 + \omega^2 = (l+n)^2 + \omega^2$$

where

$$\omega^2 = \omega_0^2 - n^2 \qquad (3.55)$$

Substituting equation (3.55) into equation (3.54), we obtain:

$$x(l) = \frac{lv_0}{(l+n)^2 + \omega^2} \qquad (3.56)$$

Before attempting to invert equation (3.56) into the time domain, note that in general $\omega^2$ may be positive, negative, or equal to zero. Let's consider all of these cases. Usually $\omega^2$ is positive and the mass performs free damped vibrations. In the other two cases, there are no vibrations and the mass performs a damped motion. Let us first consider the case when $\omega^2$ is positive.

Based on pairs 1 and 14 from Table 3.1, we invert equation (3.56) into the time domain:

$$x = \frac{v_0 e^{-nt}}{\omega} \sin \omega t \qquad (3.57)$$

Equation (3.57) provides the solution of the differential equation of motion (3.50) for the initial conditions (3.51) when $\omega^2$ is positive. Equation (3.57) shows that the motion represents damped vibrations.

When $\omega^2$ is negative, equation (3.56) obtains the following shape:

$$x(l) = \frac{lv_0}{(l+n)^2 - \omega^2} \qquad (3.56a)$$

The inversion of equation (3.56a) into the time domain is based on pairs 1 and 14a from Table 3.1 and we may write:

$$x = \frac{v_0 e^{-nt}}{\omega} \sinh \omega t \qquad (3.57a)$$

where $\sinh \omega t$ is a hyperbolic sine that according to the formula (3.a) from Table 3.1 equals:

$$\sinh \omega t = \frac{e^{\omega t} - e^{-\omega t}}{2}$$

Substituting this expression into equation (3.57a) and analyzing further, we obtain:

$$x = \frac{v_0}{2\omega}[e^{t(\omega - n)} - e^{-t(\omega + n)}]$$

This equation is the solution of the differential equation of motion (3.50) with its initial conditions (3.51) for the case when $\omega^2$ is negative. In this case, we have a damped motion without vibrations.

Finally, when $\omega^2$ equals zero, equation (3.56) will obtain the following shape:

$$x(l) = \frac{l v_0}{(l+n)^2} \tag{3.56b}$$

In order to invert equation (3.56b) into the time domain, we use pairs 1 and 21 and obtain:

$$x = v_0 t e^{-nt} \tag{3.57b}$$

This is the solution of the above-mentioned differential equation of motion, when $\omega^2$ equals zero. In this case, we also have a damped motion.

### 3.4.6   Motion of a Damped System Subjected to a Time Depending Force

Consider a body moving horizontally on a frictionless surface and experiencing a damping resistance. The motion of this body is caused by the action of an active force $Q_1$ that is increasing its value during a certain interval of time. This is the first example that actually incorporates an active force. In order to include an active force in the differential equation of motion, it is necessary to have an appropriate analytical expression of this force. The description of its characteristics is identical to the description of the third expression in the right side of equation (1.6). Thus, the analytical expression of the active force for this case has the following form:

$$Q_1 = Q(1 + \frac{\mu t}{\tau})$$

where $\tau$ is the interval of time during which the motion may last, $\mu$ is a constant dimensionless coefficient, and $\tau$ and $\mu$ have positive values.

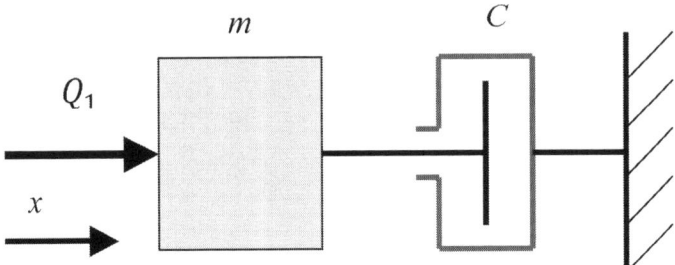

**Figure 3.4 A model of a damped system loaded by an external force.**

The model of this mechanical system is shown in Figure 3.4. Based on this model and referring to equation (1.6), we can compose the following differential equation of motion:

$$m\frac{d^2x}{dt^2} + C\frac{dx}{dt} = Q(1 + \frac{\mu t}{\tau}) \tag{3.58}$$

where $Q$ is the value of the active force at the beginning of the motion. The initial conditions of motion for the equation (3.58) may be arbitrary; however, in order to simplify the solution we accept that

$$\text{for} \quad t = 0 \quad x = 0; \quad \frac{dx}{dt} = 0 \tag{3.59}$$

Dividing equation (3.58) by $m$ and using equation (3.31) we have:

$$\frac{d^2x}{dt^2} + 2n\frac{dx}{dt} = q(1 + \frac{\mu t}{\tau}) \tag{3.60}$$

where

$$q = \frac{Q}{m}$$

Using pairs 1, 2, 3, 16, and 17 from Table 3.1, we convert equation (3.60) with its initial conditions (3.59) into the Laplace domain:

$$l^2 x(l) + 2nlx(l) = q + \frac{\mu q}{\tau l} \tag{3.61}$$

Solving equation (3.61) for the displacement in Laplace domain $x(l)$, we obtain:

$$x(l) = \frac{q}{l(l+2n)} + \frac{\mu q}{\tau l^2 (l+2n)} \tag{3.62}$$

The fractions in the right-hand side part of the equation (3.62) do not have equivalent representations in the Table 3.1, thus we need to apply to these fractions the method of decomposition. We will do this for each fraction separately. For the first fraction we may write:

$$\frac{1}{l(l+2n)} = \frac{A}{l} + \frac{B}{l+2n}$$

$$1 = A(l+2n) + Bl$$

or

$$1 = l(A+B) + 2nA$$

$$0 = A + B, \qquad 1 = 2nA,$$

$$A = \frac{1}{2n}, \quad B = -\frac{1}{2n}$$

For the second fraction we have:

$$\frac{1}{l^2(l+2n)} = \frac{D}{l} + \frac{E}{l^2} + \frac{G}{l+2n}$$

$$1 = Dl^2 + 2Dln + El + 2En + Gl^2$$

$$1 = l^2(D+G) + l(2Dn+E) + 2En$$

$$0 = D+G; \quad 0 = 2Dn+E; \quad 1 = 2En$$

$$E = \frac{1}{2n}; \quad 0 = 2Dn + \frac{1}{2n}; \quad D = -\frac{1}{4n^2}; \quad G = \frac{1}{4n^2}$$

Applying the results of the decomposition to equation (3.62) we obtain:

$$x(l) = \frac{q}{2nl} - \frac{q}{2n(l+2n)} - \frac{\mu q}{4n^2\tau l} + \frac{\mu q}{2n\tau l^2} + \frac{\mu q}{4n^2\tau(l+2n)}$$

In order to invert this equation, the denominator of the second and last fractions should be restructured, as it is shown below:

$$x(l) = \frac{q}{2nl} - \frac{q}{2n[l-(-2n)]} - \frac{\mu q}{4n^2\tau l} + \frac{\mu q}{2n\tau l^2}$$
$$+ \frac{\mu q}{4n^2\tau[l-(-2n)]} \tag{3.63}$$

Using pairs 1, 2, 3, 4, and 5 from Table 3.1, we invert equation (3.63) into the time domain:

$$x = \frac{qt}{2n} + \frac{q}{4n^2}(e^{-2nt} - 1) - \frac{\mu qt}{4n\tau} + \frac{\mu qt^2}{4n\tau}$$
$$- \frac{\mu q}{8n^3\tau}(e^{-2nt} - 1) \tag{3.64}$$

Further analyzing (3.64), we finally have:

$$x = \frac{q}{2n}\left[\left(1 - \frac{\mu}{2n\tau}\right)\left(t + \frac{e^{-2nt} - 1}{2n}\right) + \frac{\mu t^2}{2n\tau}\right] \qquad (3.65)$$

Equation (3.65) represents the solution of the differential equation of motion (3.58) with the initial conditions (3.59).

### 3.4.7 Forced Motion with Damping and Stiffness

Let's consider a mechanical system that is made up of a mass that is linked to a dashpot and a spring that are acting in parallel. The system is subjected to the action of a constant force $R$. The mass moves horizontally on a frictionless surface. Figure 3.5 shows the model of this mechanical system. Referring to equation (1.6) and Figure 3.5, we can compose the following differential equation of motion:

$$m\frac{d^2 x}{dt^2} + C\frac{dx}{dt} + Kx = R \qquad (3.66)$$

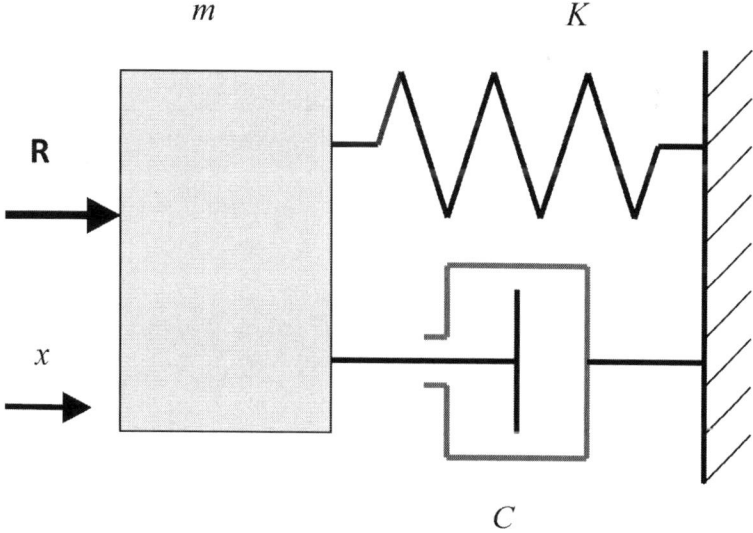

**Figure 3.5 The model of a vibratory system subjected to a constant force.**

Dividing equation (3.66) by $m$, we obtain:

$$\frac{d^2 x}{dt^2} + 2n\frac{dx}{dt} + \omega_0^2 x = r \tag{3.67}$$

where

$$r = \frac{R}{m} \tag{3.67a}$$

Based on pairs 1, 2, 16, and 17 from Table 3.1, we can convert equation (3.67) with its initial conditions (3.17) into the Laplace domain:

$$l^2 x(l) - lv_0 - l^2 s_0 + 2nlx(l) - 2nls_0 + \omega_0^2 x(l) = r$$

Solving this equation for the displacement in the Laplace domain, $x(l)$ we have:

$$x(l) = \frac{r}{l^2 + 2nl + \omega_0^2} + \frac{l(v_0 + 2ns_0)}{l^2 + 2nl + \omega_0^2}$$
$$+ \frac{l^2 s_0}{l^2 + 2nl + \omega_0^2} \tag{3.68}$$

Modifying the denominators of equation (3.68) by appropriately using equation (3.55), we obtain:

$$x(l) = \frac{r}{(l+n)^2 + \omega^2} + \frac{l(v_0 + 2ns_0)}{(l+n)^2 + \omega^2}$$
$$+ \frac{l^2 s_0}{(l+n)^2 + \omega^2} \tag{3.69}$$

Recall that the parameter $\omega^2$ can be positive, negative, or equal to zero. All three cases were considered above, making it reasonable to limit our analysis to the positive case.

Referring to pairs 1, 2, 13, 14, and 15 from Table 3.1, we invert equation (3.69) into the time domain:

$$x = \frac{r}{\omega^2 + n^2}[1 - e^{-nt}(\cos\omega t + \frac{n}{\omega}\sin\omega t)]$$

$$+ \frac{v_0 + 2ns_0}{\omega}e^{-nt}\sin\omega t \qquad\qquad (3.70)$$

$$+ s_0(\cos\omega t - \frac{n}{\omega}\sin\omega t)e^{-nt}$$

Further analyzing equation (3.70), we have:

$$x = \frac{r}{\omega^2 + n^2} - e^{-nt}\{\frac{r - s_0(\omega^2 + n^2)}{\omega^2 + n^2}\cos\omega t$$

$$-[\frac{rn - (\omega^2 + n^2)(v_0 + ns_0)}{\omega(\omega^2 + n^2)}]\sin\omega t\} \qquad (3.71)$$

Equation (3.71) may undergo more algebraic actions, but they are not needed at this time.

### 3.4.8   Forced Vibrations

Figure 3.6 illustrates a vibratory system with a mass $m$ that is connected to a spring characterized by a stiffness coefficient $K$. The mass is subjected to the action of an external sinusoidal force $A\sin\omega_1 t$, while $A$ is the force amplitude.

The mass is moving in the horizontal direction on a frictionless surface. Based on these considerations and referring to both equation (1.6) and Figure 3.6, we compose the following differential equation of motion of this system:

$$m\frac{d^2x}{dt^2} + Kx = A\sin\omega_1 t \qquad\qquad (3.72)$$

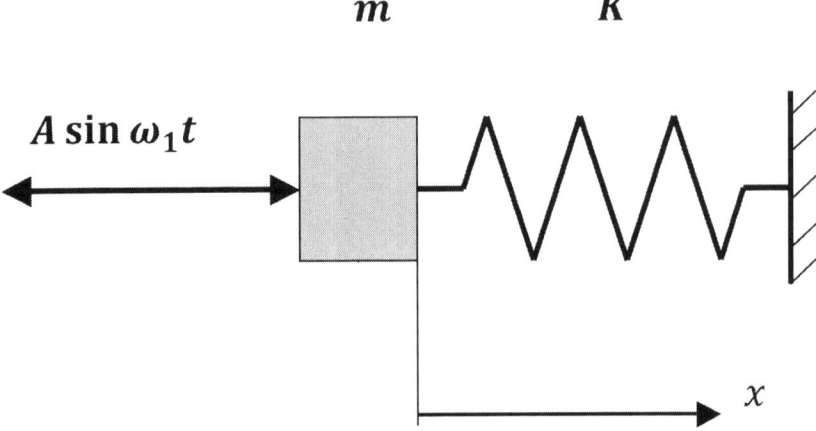

**Figure 3.6 The model of a vibratory system subjected to a sinusoidal force.**

The initial conditions of motion are according to expression (3.59):

$$\text{for} \quad t = 0 \quad x = 0; \quad \frac{dx}{dt} = 0$$

Dividing equation (3.72) by $m$, we obtain:

$$\frac{d^2x}{dt^2} + \omega_0^2 x = a \sin \omega_1 t \tag{3.73}$$

where

$$a = \frac{A}{m} \tag{3.74}$$

Applying pairs 1, 2, 8, 16, and 17 from Table 3.1, we convert equation (3.73) with its initial conditions (3.59) into the Laplace domain:

$$l^2 x(l) + \omega_0^2 x(l) = \frac{a\omega_1 l}{l^2 + \omega_1^2} \tag{3.75}$$

Solving equation (3.75) for $x(l)$, we may write:

$$x(l) = \frac{a\omega_1 l}{\left(l^2 + \omega_0^2\right)\left(l^2 + \omega_1^2\right)} \qquad (3.76)$$

Applying Table 3.1's pair 18 to equation (3.76), we invert this equation into the time domain form:

$$x = \frac{a(\omega_0 \sin \omega_1 t - \omega_1 \sin \omega_o t)}{\omega_0(\omega_0^2 + \omega_1^2)} \qquad (3.77)$$

Equation (3.77) provides the solution to the differential equation of motion (3.72) with initial conditions (3.59).

This chapter shows examples of solving differential equations of motion. Understanding these examples should be helpful when solving differential equations of motion associated with real problems.

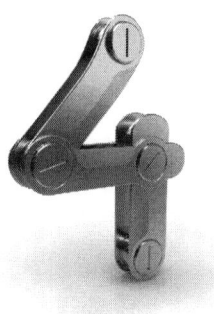

# ANALYSIS OF TYPICAL MECHANICAL ENGINEERING SYSTEMS

Numerous engineering courses contain comprehensive descriptions of certain aspects of dynamics of mechanical system. The studies of engines, automobiles, locomotives, watercraft, aircraft, handling and construction machinery, machining equipment, robotics, and more are offered on different levels in many educational enterprises. For our purposes, however, a simplified approach lets us demonstrate in a straightforward manner the step-by-step method for analyzing mechanical systems.

Normally we start by formulating the problem, focusing on factors that play decisive roles. Next, we assemble the differential equation of motion, accounting for these factors. The equation should reflect the level of the required accuracy suitable to the appropriate stage of the investigation. In many cases, not all aspects of the problem are clear before finalizing the structure of the differential equation of motion. In these situations, a few iterations of the differential equation of motion could be required.

In the next step, we solve the differential equation of motion using the Laplace Transform. Then in the last step, we analyze the equation of motion in order to find the system's performance characteristics, the role and influence of its parameters, and other related engineering factors. A simplified approach is often justifiable in order to get preliminary information that lets us estimate the values of the important parameters. Then we can decide about the appropriate venues for continuing the investigation.

We analyze a mechanical system's dynamics to gather information that will help us develop and improve that system. When we know the system's parameters of motion, our analysis can reveal information about required power, energy consumption, performance, applied forces, level of acceleration/deceleration, etc. The information about power helps us select an energy source; the information about the force is used for stress calculations. The level of acceleration/deceleration and frequency of vibrations are associated with human health and safety issues. The performance and energy consumption reflect the efficiency and productivity of the system.

When new mechanical systems are being developed, their performance parameters should be well thought-out as part of the design process. This analysis leads to many engineering calculation methodologies that result in successful solutions. The following examples illustrate these methodologies. The quality of the analysis depends on the level of the accuracy by which the differential equation of motion reflects the real characteristics of the problem.

The solution of the differential equation of motion represents the displacement of the system's mass as a function of time. Taking the first derivative from the displacement, we obtain the velocity of the mass as a function of time. Taking the second derivative, we get the expression for the acceleration as a function of time.

## 4.1  Lifting a Load

Suppose you are developing a mechanical system for lifting a load of materials. The lifting mechanism should incorporate an engine coupled with a winch that includes a braking system and means of control. The load should be attached to a steel cable, which, by the help of a corresponding system of pulleys, forms a closed loop that is linked to the winch. A braking mechanism is coupled with the winch; this mechanism is powered by an electrical motor through a gearbox. For maintenance considerations, the mechanism is mounted on the floor. Assume the friction and air resistance are negligible during the process of the motion of the load. The weight of the load and the velocity of the uniform upward motion are pre-determined by industrial norms and regulations.

The purpose of the analysis is to determine the system's power needs and the magnitude of the forces applied to the system, and also to evaluate the performance characteristics. Knowing the power requirements helps us select the appropriate source of energy, whereas the information about forces is needed for the stress calculations of the cable and other links. The analysis usually includes several options of performance characteristics; by comparing their results, we can select the most appropriate option.

The lifting process comprises the following three steps: acceleration, uniform motion, and deceleration or braking. The displacement during uniform motion is calculated by multiplying velocity by running time; it does not require a specific analysis. There are no readily available formulas to calculate the process of acceleration and braking (deceleration). These processes can be described by solving appropriate differential equations of motion, as presented below.

### 4.1.1  Acceleration

Figure 4.1 shows a model of lifting, where $W$ is the weight of the load, and $R$ is the constant active force representing the lifting force. The axis $x$ is directed upward. Based on Figure 4.1 and equation (1.6), we can assemble the following differential equation of motion for this case:

$$m\frac{d^2t}{dt^2} + W = R \tag{4.1}$$

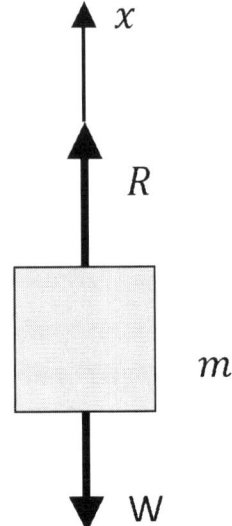

**Figure 4.1  A model of lifting.**

The initial conditions are:

$$\text{for} \quad t = 0 \quad x = 0, \quad \frac{dx}{dt} = 0 \tag{4.2}$$

Dividing equation (4.1) by $m$, we have:

$$\frac{d^2t}{dt^2} + g = r \tag{4.3}$$

where $g$ is the acceleration of gravity:

$$g = \frac{W}{m} \tag{4.4}$$

and

$$r = \frac{R}{m} \tag{4.5}$$

Many examples of using Laplace Transform pairs from Table 3.1 for conversion and inversion were provided in Chapter 3. From now on, we will specifically refer to this Table only in selected examples.

The conversion of equation (4.3) with its initial conditions (4.2) into the Laplace domain reads:

$$l^2 x(l) + g = r$$

The solution of this equation for the Laplace domain displacement $x(l)$ is:

$$x(l) = \frac{r - g}{l^2} \tag{4.6}$$

The inversion of equation (4.6) into the time domain is:

$$x = \frac{r - g}{2} t^2 \tag{4.7}$$

Differentiating equation (4.7), we obtain the expression for the velocity:

$$\frac{dx}{dt} = (r - g)t \tag{4.8}$$

In turn, differentiating equation (4.8), we determine the acceleration:

$$\frac{d^2 x}{dt^2} = r - g \tag{4.9}$$

There are two ways in which the acceleration process can be realized. The first option is to achieve the predetermined velocity at the end of a certain period of time. The second option is to get this velocity at the end of a certain displacement. Let's consider both options, keeping in mind that some parameters of the first option will be given the subscript 1, whereas these parameters for the second option will get the subscript 2.

According to the first option, at the end of time $t_1$, the velocity equals $v$. Substituting these values into equation (4.8), we write:

$$v = (r_1 - g)t_1 \tag{4.10}$$

From equation (4.10), we calculate the parameter $r_1$:

$$r_1 = \frac{v}{t_1} + g \tag{4.11}$$

Combining equation (4.11) with equations (4.4) and (4.5), we obtain:

$$R_1 = \frac{mv}{t_1} + W \tag{4.12}$$

Equation (4.12) characterizes the external force applied to the system according to the first option. Because the power is the product of multiplying the force by the velocity, we determine the required power $N_1$ for the first option:

$$N_1 = \frac{mv^2}{t_1} + Wv \tag{4.13}$$

Combining equations (4.9) and (4.11), we define the acceleration $a_1$ for the first option:

$$a_1 = \frac{v}{t_1}.$$

The displacement $s_1$ for the first option is determined by solving simultaneously equations (4.7), (4.10), and (4.11):

$$s_1 = \frac{1}{2}\, vt_1 \tag{4.14}$$

Multiplying the force from equation (4.12) by the displacement from equation (4.14), we obtain the energy consumption $E_1$ during the acceleration for the first option:

$$E_1 = \frac{1}{2}(mv^2 + Wvt_1) \qquad \textbf{(4.14a)}$$

This concludes the analysis of the first option.

According to the second option, by the end of displacement $s_2$, the mass will obtain the velocity $v$. Equating the left side of equation (4.7) to $s_2$, we determine the time duration $t_2$ for the second option:

$$t_2 = \sqrt{\frac{2s_2}{r_2 - g}} \qquad \textbf{(4.15)}$$

Setting the left side of equation (4.8) equal to $v$ and substituting into it the time according to equation (4.15), we solve for the force $R_2$ that acts during the second option:

$$R_2 = \frac{mv^2}{2s_2} + W \qquad \textbf{(4.16)}$$

The required power $N_2$ is calculated by multiplying the force according to equation (4.16) by the velocity:

$$N_2 = \left(\frac{mv^2}{2s_2} + W\right)v \qquad \textbf{(4.17)}$$

The energy consumption $E_2$ for the second option equals the product of multiplying the force according to equation (4.16) by the displacement $s_2$:

$$E_2 = \frac{mv^2}{2} + Ws_2 \qquad \textbf{(4.17a)}$$

In order to compare the energy consumptions of these two options, we subtract equation (4.17a) from equation (4.14a), leading to:

$$\Delta E = E_1 - E_2 = W(\frac{vt_1}{2} - s_2)$$

where $\Delta E$ is the difference of the consumed energies.

The values of the parameters $t_1$ and $s_2$ are chosen based on certain practical considerations. If we obtain that $\Delta E > 0$ when we substitute these parameters, the first option is more energy consuming than the second.

Subtracting equation (4.16) from equation (4.12), we can compare the acting forces for these options:

$$\Delta R = R_1 - R_2 = mv\left(\frac{1}{t_1} - \frac{v}{2s_2}\right)$$

where $\Delta R$ is the difference between the forces; if it is positive, then the first option needs a stronger cable and more power than the second.

Based on this comparison, it becomes possible to select the most appropriate option.

### 4.1.2 Braking

The braking mechanism is intended to bring the load to a stop in a predetermined point and hold it at this point. The braking process starts at a certain moment before the load reaches its destination. At the same moment, the lifting force is removed while the load continues to move upward by inertia (by its kinetic energy). If the active force does not act, the load will be stopped just by its gravity force, without any involvement of a braking mechanism. However, if a braking mechanism is needed to decrease the braking distance or time, it should be applied and activated at the beginning of the deceleration process.

Usually the braking mechanism applies a certain friction force, which acts together with the gravity force. At the point where the load stops, the frictional braking force instantaneously becomes directed upward; it should prevent the free fall of the load. In order to hold the load in the standstill position, the braking force should equal at least the force of the load's gravity. In this case, during deceleration, the mass is subjected to a total resisting force that will exceed at least two times the force of the load's gravity.

In turn, we should calculate the braking distance, which will let us determine the moment a certain control unit should be activated, causing the interruption of the lifting force and engaging the brake.

We express the breaking friction force applied to the mass in terms of the force of gravity, namely: $\varepsilon W$, where $\varepsilon$ is a coefficient that characterizes the applied friction force. If the braking is caused just by the gravity, then $\varepsilon = 0$. If friction is added to the weight, then $\varepsilon \geq 1$. At the beginning of the braking process, the energy source is disconnected; the upward motion continues due to the kinetic energy of the load that has a velocity $v$. As with the lifting process, the air resistance during the braking is ignored. Factoring these considerations and referring to equation (1.1), we write the following differential equation of motion for the braking process:

$$m\frac{d^2x}{dt^2} + W(1+\varepsilon) = 0 \qquad (4.18)$$

At the end of the braking process, the load should be stopped at a predetermined height. For the analysis of the motion, the load's position at the start of the braking process does not matter. Thus, we move the origin of the coordinate system to the point where the braking process starts. The initial conditions of motion for equation (4.18) are:

$$\text{for} \quad t = 0 \quad x = 0; \quad \frac{dx}{dt} = v \qquad (4.19)$$

Dividing equation (4.18) by $m$, we have:

$$\frac{d^2x}{dt^2} + g(\varepsilon + 1) = 0 \qquad (4.20)$$

The Laplace domain of equation (4.20) at the initial conditions (4.19) reads:

$$l^2 x(l) - lv + g(\varepsilon + 1) = 0 \qquad (4.21)$$

The Laplace domain solution of equation (4.21) is:

$$x(l) = \frac{v}{l} - \frac{g(\varepsilon + 1)}{l^2} \qquad (4.22)$$

The inversion of equation (4.22) into the time domain reads:

$$x = vt - \frac{1}{2} g t^2 (\varepsilon + 1) \qquad (4.23)$$

Taking the first derivative from equation (4.23), we obtain the expression for the velocity:

$$\frac{dx}{dt} = v - gt(\varepsilon + 1) \qquad (4.24)$$

Keeping in mind $\varepsilon = 0$ in cases when the braking is achieved just by the gravity force, and equating the left side of equation (4.24) to zero, we determine the braking time $t_0$ for this case:

$$t_0 = \frac{v}{g}$$

Substituting this time into equation (4.23), we calculate the braking distance $s_0$ for this case:

$$s_0 = \frac{v^2}{2g}$$

The braking time $t_0$ and braking distance $s_0$ have the largest values for the braking process based just on the force of gravity. By adding a friction force to the force of gravity, we can reduce these values. We have to analyze two options. The first option is when the braking time $t_1$ should be less than $t_0$. For this option, the coefficient $\varepsilon$ is given a subscript 1, and we write $\varepsilon_1 \geq 1$. At the end of the period of time $t_1$ the velocity should equal zero. Setting the left side of equation (4.24) equal to zero and substituting into it time $t_1$, we calculate the required braking force ($\varepsilon_1 + 1)W$ and the value of $\varepsilon_1$ for the first option:

$$(\varepsilon_1 + 1)W = \frac{mv}{t_1}$$

$$\varepsilon_1 = \frac{mv}{t_1 w} - 1 \qquad \textbf{(4.24a)}$$

Substituting $t_1$ and $\varepsilon_1$ into equation (4.23), we determine the braking distance $s_1$ for the first option:

$$s_1 = vt_1 - \frac{1}{2} gt_1^2 (\varepsilon_1 + 1)$$

For the second option, we consider that the braking distance $s_2$ is less than $s_0$. Replacing the coefficient $\varepsilon_1$ by $\varepsilon_2$ and setting the velocity in equation (4.24) equal to zero, we determine the time $t_2$ that the process lasts during this option:

$$t_2 = \frac{v}{g(\varepsilon_2 + 1)} \qquad \textbf{(4.25)}$$

Equating the left side of equation (4.23) to $s_2$ and substituting into it the time according to equation (4.25), we determine the value of the coefficient $\varepsilon_2$ for the second option:

$$\varepsilon_2 = \frac{v^2}{2s_2 g} - 1 \qquad \textbf{(4.26)}$$

Comparing the values calculated according to equations (4.24a) and (4.26), we make the final decision regarding the preferable option.

## 4.2  Water Vessel Dynamics

Let's analyze the acceleration process of a ship: our goal is to determine the power needed to achieve a certain velocity. For this example, the influences of the wind and water stream activities are not considered. In addition, the decrease of the ship's mass due to fuel consumption is ignored.

A vessel in motion experiences simultaneously a damping resistance of the water and the air. Figure 4.2 shows the model of a vessel in motion, where $R$ is a constant active force developed by the source of power and $C_1$ and $C_2$ are the damping coefficients of air and water respectively. Assumed both damping coefficients are constant.

According to equation (1.7), the resultant damping coefficient is:

$$\dot{C} = C_1 + C_2 \qquad\qquad (4.27)$$

Building on Figure 4.2 and equations (1.6) and (4.27), we can develop the following differential equation of motion:

$$m\frac{d^2t}{dt^2} + C\frac{dx}{dt} = R \qquad\qquad (4.28)$$

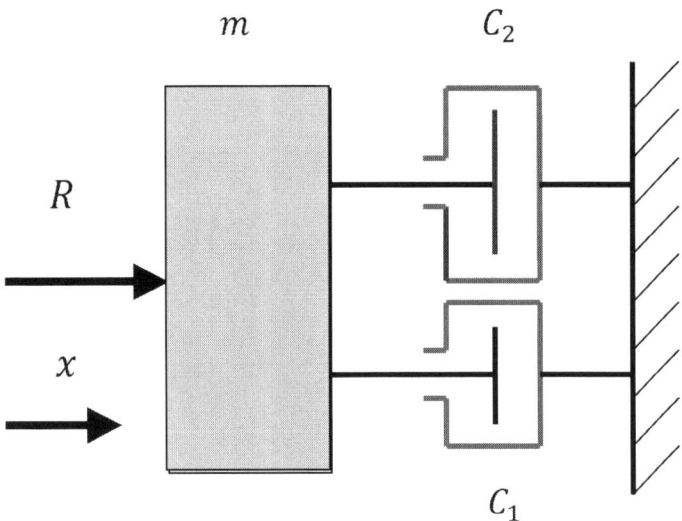

**Figure 4.2  The model of a ship in motion.**

The initial conditions are:

$$\text{for} \quad t = 0 \quad x = 0; \quad \frac{dx}{dt} = 0 \qquad \textbf{(4.29)}$$

Dividing equation (4.28) by $m$, and recalling equations (3.31) and (3.67), we have:

$$\frac{d^2x}{dt^2} + 2n\frac{dx}{dt} = r \qquad \textbf{(4.30)}$$

The Laplace domain of equation (4.30) at the initial conditions (4.29) reads:

$$l^2x(l) + 2nlx(l) = r \qquad \textbf{(4.31)}$$

The solution of equation (4.31) in the Laplace domain is:

$$x(l) = \frac{r}{l(l + 2n)} \qquad \textbf{(4.32)}$$

The right side of equation (4.32) does not have a representation in Table 3.1. However, a similar expression and its decomposition are considered in section 3.4.6 with regard to equation (3.62); that contains a parameter $q$ instead of the parameter $r$ in the numerator. Thus, based on equation (3.63), equation (4.32) is transformed as follows:

$$x(l) = \frac{r}{2nl} - \frac{r}{2n[l - (-2n)]} \qquad \textbf{(4.33)}$$

The inversion of equation (4.33) into the time domain reads:

$$x = \frac{r}{2n}(t + \frac{e^{-2nt} - 1}{2n}) \qquad \textbf{(4.34)}$$

Differentiating equation (4.34), we obtain the expression for the velocity:

$$\frac{dx}{dt} = \frac{r}{2n}\left(1 - e^{-2nt}\right) \qquad (4.35)$$

Taking the first derivative from equation (4.35), we determine the acceleration:

$$\frac{d^2x}{dt^2} = re^{-2nt} \qquad (4.36)$$

Equation (4.36) shows that, at the beginning of the motion, the acceleration has its maximum value and then it decreases with time. Thus, the maximum acceleration $a_{max}$ equals:

$$a_{max} = r \qquad (4.37)$$

Combining equations (4.37) and (3.67) we have:

$$a_{max} = \frac{R}{m} \qquad (4.38)$$

According to equation (4.38), the value of the maximum acceleration should comply with the corresponding values of public health and safety regulations related to public transportation. In general, velocity reaches its maximum value when acceleration becomes equal to zero. The limit of the exponential function in the right side of equation (4.36) becomes equal to zero when time tends to an infinite value, while $r$ is a constant value. In reality, acceleration cannot equal zero; consequently, velocity will never reach its maximum value. However, you may consider the condition at which acceleration tends to zero. Based on equation (4.36), this condition can be expressed as follows:

$$re^{-2nt_1} \to 0 \qquad (4.39)$$

where $t_1$ is the time duration at which acceleration approaches zero and velocity tends to its maximum value. Obviously, $t_1 \to \infty$. According to equation (4.39), we determine the condition at which velocity approaches its maximum value:

$$e^{-2nt_1} \to 0 \qquad \textbf{(4.40)}$$

Combining expression (4.40) with equation (4.35), we calculate the maximum velocity $v_{max}$ that the ship can approach:

$$v_{max} \to \frac{r}{2n} \qquad \textbf{(4.41)}$$

Coordinating equations (3.31) and (3.67) with expression (4.41), we have:

$$v_{max} \to \frac{R}{C} \qquad \textbf{(4.42)}$$

The required power $N$ is the product of multiplying the applied force by the maximum velocity:

$$N = \frac{R^2}{C} \qquad \textbf{(4.43)}$$

Equation (4.43) provides the information needed to find the vessel's required energy source.

Similar equations, solutions, and results could be obtained by analyzing the acceleration of an aircraft moving through the air in a horizontal direction. In that case, we assume the damping coefficient of the air is a constant value, and both the influence of the wind and decrease of the mass due to fuel consumption are insignificant.

## 4.3  Dynamics of an Automobile

The process of motion for an automobile or other wheeled ground transportation consists of three phases: acceleration, uniform motion, and deceleration (or braking). The analysis of the automobile dynamics is focused on the phases of acceleration and deceleration. The uniform motion is characterized by a constant velocity, and the displacement of the automobile represents a linear function of time. This is well known from basic physics. However, because of fuel consumption, the total mass of the automobile is gradually decreasing. This imposes certain complications on the analysis of the uniform motion. We will investigate just the processes of acceleration and deceleration — and assume that the change of the mass is insignificant. To simplify our analysis, we will consider the motion of an automobile in the horizontal direction and assume that the damping coefficient of the air resistance has a constant value.

### 4.3.1 Acceleration

Our goals are to determine the power required by the energy source to accelerate an automobile to its maximum velocity, then to reveal the relationship between the parameters of the system. In addition, we want to estimate the level of the maximum acceleration with respect to public health and safety regulations.

Let's consider the dynamics of a rear-wheel drive vehicle in the process of accelerating. Figure 4.3 shows the model of a rear-wheel drive automobile.

As Figure 4.3 shows, the weight $W$ of the automobile and its force of inertia $IF = m\frac{d^2x}{dt^2}$ are applied to the center of gravity G of the system. A torque $T$ produced by the engine is transmitted to the rear axle that rotates the rear wheels around point $O_1$. The torque equals:

$$T = R\frac{D}{2} \tag{4.44}$$

where $R$ is the tangential force applied to the point A of the wheels where they touch the ground, whereas $D$ is the diameter of the wheel. Point A is concurrently subjected to the action of four forces. Two equal and opposite directed forces $W_A$ and $Y_A$ act in the

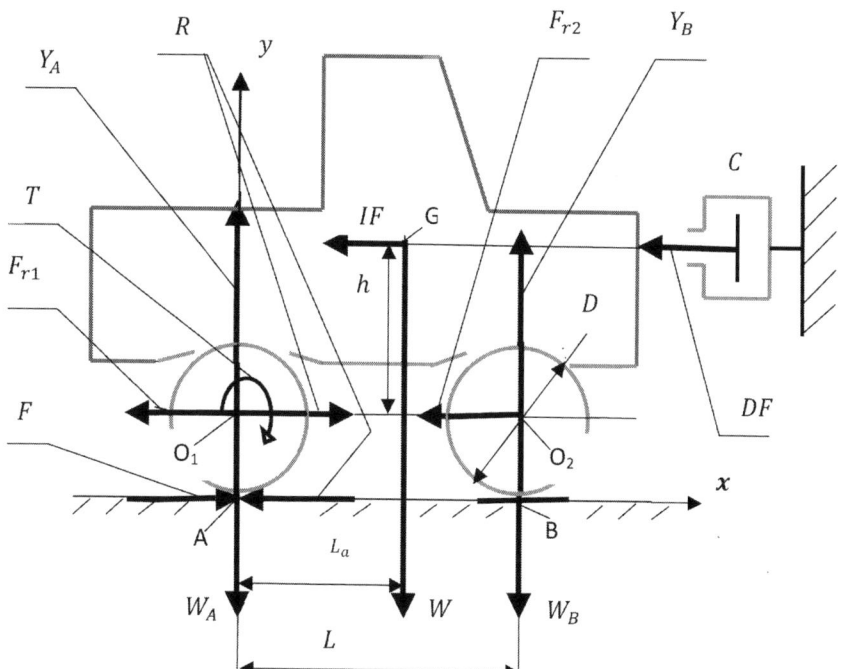

**Figure 4.3 The model of a rear-wheel drive automobile.**

vertical direction while two equal and opposite directed forces $F$ and $R$ act in the horizontal direction. The weight of the automobile is distributed between the rear and front wheels. The force $W_A$ represents the partial load of the automobile on the rear wheels; it is transmitted to the ground. The force $Y_A$ is the reaction of the ground to the force $W_A$; it is applied to the wheels.

The horizontal forces are the tangential force $R$ and the dry friction force $F$. The point A is the instantaneous center of rotation of the rear wheels; if the wheels are not slipping, the velocity of the wheels in the point A equals zero. The frictional force $F$ is exerted by the ground due to the action of the tangential force $R$ that is tending to push to the left the point A that belongs to the wheel. The friction force, however, prevents the motion of the wheel to the left of point A. All this generates a force that is equal and opposite to force $R$ and is applied to the center of the rear axle at point $O_1$.

This force causes the wheel to roll to the right, resulting in the automobile's acceleration. The front wheels touch the ground at point B in which two equal and opposite directed forces are applied. The force $W_B$ is the partial load of the automobile on the front wheels; it is transmitted to the ground while force $Y_B$ is the reaction of the ground applied to the wheels. During the process of acceleration, the force of inertia causes a redistribution of the automobile's weight between the axels so that the rear axle obtains an additional part of the weight that is subtracted from the front axle.

The rolling friction forces $F_{r1}$ and $F_{r2}$ are respectively applied to the rear axle at point $O_1$ and to the front axle at point $O_2$. These forces can be determined from the following expressions:

$$F_{r1} = f_r Y_A \qquad\qquad (4.45)$$

and

$$F_{r2} = f_r Y_B \qquad\qquad (4.46)$$

where $f_r$ is the rolling friction coefficient.

The air resistance force is distributed over the cross-sectional area of the automobile. Assume that an equivalent concentrated damping force $DF = C\frac{dx}{dt}$ is applied along the line passing through the center of gravity $G$, as is shown in Figure 4.3.

Let's look at the relationships between some of the forces. For example, the friction force equals:

$$F = \mu Y_A \qquad\qquad (4.47)$$

where $\mu$ is the friction coefficient between the wheels and the road.

The tangential force $R$ cannot exceed the friction force; consequently, the maximum value of this force equals the friction force $F$. The maximum force that can be applied to the automobile in this case is:

$$R = \mu Y_A \qquad\qquad (4.48)$$

In order to clarify the other relationships between the considered forces, we must first determine the reactions $Y_A$ and $Y_B$. In order to calculate the reactions, we will use the well-known laws of Statics. Starting with the rule that the sum of moments around any point should equal zero, we consider the corresponding sum of moments around point B:

$$\sum M_B = 0$$

The left side of the following expression represents the sum of moments around point $B$:

$$Y_A L + R \frac{D}{2} - \left(F_{r1} + F_{r2}\right) \frac{D}{2} - m \frac{d^2 x}{dt^2} \left(h + \frac{D}{2}\right)$$
$$- W\left(L - L_a\right) - C \frac{dx}{dt}\left(h + \frac{D}{2}\right) = 0 \tag{4.49}$$

where $L$ is the wheelbase, $h$ is the distance between the center of gravity and the centers of the wheels, and $L_a$ is the distance between the rear axle and the center of gravity. The values of the structural parameters of the automobile $L$, $L_a$, and $h$ are predetermined by the design considerations and become obvious from Figure 4.3.

Combining equations (4.45), (4.46), and (4.48) with equation (4.49), we have:

$$Y_A L + \mu Y_A \frac{D}{2} - \left(f_r Y_A + f_r Y_B\right) \frac{D}{2} - m \frac{d^2 x}{dt^2} \left(h + \frac{D}{2}\right)$$
$$- W\left(L - L_a\right) - C \frac{dx}{dt}\left(h + \frac{D}{2}\right) = 0 \tag{4.50}$$

This equation has two unknowns, namely the reactions $Y_A$ and $Y_B$. The additional equation that contains these two reactions is the equation of Statics; it indicates that the sum of projections of the applied forces onto the axis $y$ equals zero, so we may write:

$$\sum F_y = 0$$

According to this equation we have:

$$Y_A - W + Y_B = 0$$

From this equation, we determine:

$$Y_B = W - Y_A$$

Combining this equation with equation (4.50), we eliminate the unknown reaction $Y_B$ :

$$Y_A L + \mu Y_A \frac{D}{2} - \left(f_r Y_A + f_r W - f_r Y_A\right)\frac{D}{2} - m\frac{d^2 x}{dt^2}\left(h + \frac{D}{2}\right)$$

$$- W\left(L - L_a\right) - C\frac{dx}{dt}\left(h + \frac{D}{2}\right) = 0 \qquad (4.51)$$

The third expression of equation (4.51) shows that the total rolling friction resistance equals $f_r W$ and does not depend on the distribution of the weight between the axels. Considering this, we solve equation (4.51) for the reaction $Y_A$ :

$$Y_A = W\frac{0.5Df_r + L - L_a}{L + 0{,}5\mu D} + \frac{h + 0.5D}{L + 0.5\mu D}m\frac{d^2 x}{dt^2}$$

$$+ \frac{h + 0.5D}{L + 0{,}5\mu D}C\frac{dx}{dt} \qquad (4.52)$$

Equation (4.52) is too cumbersome to be used in further analytical procedures. Because the fractional expressions have constant values, we may accept the following notations:

$$\eta = \frac{0.5Df_r + L - L_a}{L + 0{,}5\mu D} \qquad (4.53)$$

$$\rho = \frac{h + 0.5D}{L + 0.5\mu D} \qquad (4.54)$$

Working equations (4.53) and (4.54) into equation (4.52), and factoring in equation (4.48), we write:

$$R = \mu Y_A = \mu(\eta W + \rho m \frac{d^2 x}{dt^2} + \rho C \frac{dx}{dt}) \tag{4.55}$$

Actually, as equation (4.55) indicates, the active force $R$ is a variable value.

Based on the model in Figure 4.3 and equation (1.6), we can assemble the following differential equation of motion for the automobile:

$$m \frac{d^2 x}{dt^2} + C \frac{dx}{dt} + F_{r1} + F_{r2} = R \tag{4.56}$$

Next, we replace in equation (4.56) the two components of rolling friction by the equivalent force $f_r W$, then substitute into the equation the active force according to equation (4.55). We then obtain:

$$m \frac{d^2 x}{dt^2} + C \frac{dx}{dt} + f_r W = \mu \left( \eta W + \rho m \frac{d^2 x}{dt^2} + \rho C \frac{dx}{dt} \right) \tag{4.57}$$

We can now transform equation (4.57):

$$(1 - \mu\rho) m \frac{d^2 x}{dt^2} + (1 - \mu\rho) C \frac{dx}{dt} = W(\mu\eta - f_r) \tag{4.58}$$

Analyzing the value of the dimensionless characteristic $\rho$, it can be seen from equation (4.54) that $\rho < 1$. Also, the value of friction coefficient $\mu < 1$. Obviously $\mu\rho < 1$; hence the multiplier $1 - \mu\rho > 0$. This must be emphasized because it clarifies that the force of inertia in equation (4.58) remains positive.

Dividing equation (4.58) by the factor $(1 - \mu\rho)m$, we obtain:

$$\frac{d^2 x}{dt^2} + \frac{C}{m} \frac{dx}{dt} = \frac{W}{m} (\frac{\mu\eta - f_r}{1 - \mu\rho}) \tag{4.59}$$

Let us denote:

$$2n_1 = \frac{C}{m} \tag{4.60}$$

and

$$\vartheta = \frac{\mu\eta - f_r}{1 - \mu\rho} \qquad (4.61)$$

where $n_1$ is the damping factor. Substituting equations (4.4), (4.60), and (4.61) into equation (4.59), we write:

$$\frac{d^2x}{dt^2} + 2n_1 \frac{dx}{dt} = \vartheta g \qquad (4.62)$$

The initial conditions of motion for this case are:

$$\text{for} \quad t = 0 \quad x = 0, \quad \frac{dx}{dt} = 0 \qquad (4.63)$$

Converting equation (4.62) into the Laplace domain with the initial conditions (4.63) reads:

$$l^2x(l) + 2n_1 lx(l) = \vartheta g \qquad (4.64)$$

The solution of equation (4.64) in the Laplace domain is:

$$x(l) = \frac{\vartheta g}{l(l + 2n_1)} \qquad (4.65)$$

Equation (4.65) is similar to equation (4.32). Consequently, the inversion of equation (4.65) into the time domain is similar to equation (4.34). Thus, we may write:

$$x = \frac{\vartheta g}{2n_1}(t + \frac{e^{-2n_1 t} - 1}{2n_1}) \qquad (4.66)$$

Taking the first derivative from equation (4.66), we obtain the velocity:

$$\frac{dx}{dt} = \frac{\vartheta g}{2n_1}(1 - e^{-2n_1 t}) \qquad (4.67)$$

One of an automobile's performance characteristics is the value of the velocity that can be achieved after a certain period of time from the start point. Substituting different values of time into equation (4.67) and estimating the coefficient $\vartheta$ according to equation (4.61), we can calculate the corresponding values of the velocity.

Differentiating equation (4.67), we determine the acceleration:

$$\frac{d^2x}{dt^2} = \vartheta g e^{-2n_1 t} \tag{4.68}$$

The maximum velocity is achieved when acceleration equals zero. Equating the left side of equation (4.68) to zero, we find that acceleration approaches zero when:

$$e^{-2n_1 t_1} \to 0 \tag{4.69}$$

where $t_1$ is the duration in time of the acceleration process. Combining equation (4.67) and expression (4.69), we determine the maximum velocity:

$$V_{max} \to \frac{\vartheta g}{2n_1} \tag{4.70}$$

Equation (4.55) shows that the force $R$ is changing its value during the process of acceleration. At the beginning of the acceleration, at $t = 0$, the velocity equals zero, and the acceleration according to equation (4.68) equals $\vartheta g$. Substituting these values of velocity and acceleration into equation (4.55), we calculate the value of the active force $R_0$ at the beginning of the acceleration:

$$R_0 = \mu \eta W + \rho \vartheta m g = W(\mu \eta + \rho \vartheta) \tag{4.71}$$

At the end of the acceleration process, acceleration tends to zero and velocity approaches its maximum value according to expression (4.71). Substituting these values into equation (4.55), we determine the value of the active force $R_1$ at the end of the acceleration process:

$$R_1 = \mu \eta W + \psi C \frac{\vartheta g}{2n_1}$$

Combining this equation with equation (4.60), we have:

$$R_1 = W[\mu\eta + \frac{\psi(1-\mu\rho)}{1-\mu\psi}]$$ (4.72)

For conventional automobiles, $R_0$ exceeds $R_1$; therefore, in order to determine the required power $N$ of the engine, we multiply the values of equation (4.71) and expression (4.70):

$$N = W\frac{\vartheta g}{2n_1}(\mu\eta + \rho\vartheta)$$ (4.73)

The required torque equals:

$$T = R_0 \frac{D}{2}$$ (4.74)

The maximum angular velocity of the wheels is:

$$\omega_{max} = \frac{2v_{max}}{D} = \frac{\vartheta g}{Dn_1}$$ (4.75)

According to equation (4.75), the maximum number of revolutions per minute of the wheels is:

$$n_{max} = \frac{30\omega_{max}}{\pi}$$ (4.76)

Multiplying the velocity ratio $i_t$ of the transmission between the engine and the rear axle by the maximum number of revolutions per minute of the rear axle according to the expression (4.76), we can determine the number of revolutions per minute of the engine $n_{eng}$:

$$n_{eng} = i_t n_{max}$$ (4.77)

Knowing the required power, torque, and angular velocity lets us select an appropriate engine for the automobile.

According to equation (4.68), the maximum acceleration $a_{max}$ occurs at the very beginning of the acceleration process; it is calculated from the following expression:

$$a_{max} = \vartheta g$$ (4.78)

Analyzing equation (4.61), it can be seen that $\vartheta < 1$ for conventional automobiles, which means that the maximum acceleration is less than the acceleration of gravity. However, for real life problems, the maximum acceleration should comply with the requirements and regulations pertaining to public health and safety.

## 4.3.2  Braking

From this investigation, we can determine the automobile's braking distance, and also evaluate the maximum deceleration, keeping in mind that it should be in compliance with the requirements of public health and safety.

Let's consider the braking process of an automobile having all four wheels locked. In this case, because the wheels are slipping, the road surface exerts a frictional resisting force. The redistribution of the weight between the axels does not change the total weight of the vehicle. Therefore, the resisting friction force equals:

$$F = \mu W \qquad (4.79)$$

In addition to the frictional resistance, we will consider also the damping resistance of the air. The automobile moves by inertia, possessing a certain velocity at the beginning of braking process. Thus, referring to equation (1.1), we can compose the following differential equation of motion for braking:

$$m\frac{d^2x}{dt^2} + C\frac{dx}{dt} + F = 0 \qquad (4.80)$$

The initial conditions are:

$$\text{for} \quad t = 0 \quad x = 0; \quad \frac{dx}{dt} = v_0 \qquad (4.81)$$

Dividing the equation (4.80) by $m$, we have:

$$\frac{d^2x}{dt^2} + 2n\frac{dx}{dt} + f = 0 \qquad (4.82)$$

Converting equation (4.82) into the Laplace domain reads:

$$l^2 x(l) - l v_0 + 2nl x(l) + f = 0 \tag{4.83}$$

The solution of equation (4.83) in the Laplace domain is:

$$x(l) = \frac{v_0}{l + 2n} - \frac{f}{l(l + 2n)} \tag{4.84}$$

The inversion of equation (4.84) into the time domain is similar to the inversion of the equations considered before in section 3.4.6, and it reads:

$$x = \frac{1}{2n}\left[ v_0\left(1 - e^{-2nt}\right) - f\left(t + \frac{e^{-2nt} - 1}{2n}\right) \right] \tag{4.85}$$

Taking the first derivative from the function (4.85) we determine the velocity:

$$\frac{dx}{dt} = (v_0 + \frac{f}{2n})e^{-2nt} - \frac{f}{2n} \tag{4.86}$$

The second derivative from equation (4.85) reveals the deceleration:

$$\frac{d^2 x}{dt^2} = (2n v_0 + f)e^{-2nt} \tag{4.87}$$

Equating the left side of equation (4.86) to zero, we determine the time $t_2$ the braking process takes place:

$$e^{-2n t_2} = \frac{f}{2n v_0 + f} \tag{4.88}$$

and

$$t_2 = \frac{1}{2n} \ln\left( \frac{2n v_0 + f}{f} \right) \tag{4.89}$$

Combining equations (4.85), (4.88), and (4.89), we determine the braking distance $s_2$.

$$s_2 = \frac{1}{2n}\left[\left(v + \frac{f}{2n}\right)\left(1 - \frac{f}{2nv_0 + f}\right) - \frac{f}{2n}\ln\left(\frac{2nv_0 + f}{f}\right)\right] \quad (4.90)$$

Analyzing equation (4.87), we find that the maximum deceleration $a_2$ occurs at the beginning of the braking process and it equals:

$$a_2 = 2nv_0 + f = \frac{g}{W}(Cv_0 + F) \quad (4.91)$$

The level of deceleration should comply with the regulations and requirements of public health and safety.

## 4.4    Acceleration of a Projectile in the Barrel

In this section, our main goal is to reveal the relationships between the parameters of the system. A projectile is accelerated by the pressure force of the gases from the explosive material. The dry friction force that is exerted between the barrel and the projectile, as well as the air in front of the projectile, resist its motion. Thus, the projectile is subjected to the action of the active pressure force and the resisting forces of both friction and damping. Let's analyze these forces.

The pressure forces applied to both the projectile and the barrel cause the projectile to move forward and the barrel to move backward. As a result, the volume of the rear chamber behind the projectile increases, leading to a decrease of the pressure forces in the rear chamber. Consequently, the active force applied to the projectile decreases during the projectile's acceleration. To simplify the analysis, we accept the barrel's displacement as insignificant compared with the projectile's displacement; we also accept the decrease of pressure in the rear chamber as proportional to the projectile's displacement. Furthermore, we assume the projectile moves horizontally, and that the damping coefficient $C$ of the air resistance has a constant value. Figure 4.4 illustrates a barrel-projectile system.

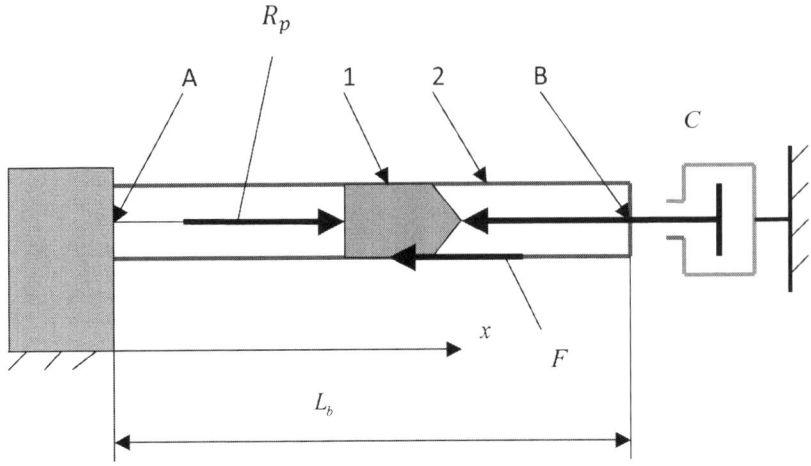

**Figure 4.4    Barrel-projectile system.**

Figure 4.4 shows that the pressure force $R_p$ causes the motion of the projectile 1 along the axis $x$. The friction force $F$ (between the barrel 2 and the projectile 1) and the damping force resist the motion of the projectile. The damping force is represented in Figure 4.4 by the dashpot, which is characterized by the damping coefficient $C$. The maximum value of the pressure force $R$ is applied to the projectile at the moment of the motion's beginning — when its tail is at point A. At point B, where the projectile leaves the barrel, the pressure force gets its minimum value. Because we consider that the pressure force decreases proportionally to the projectile's displacement, we may write the following expression that reflects the dependence of the active force $R_p$ on the projectile's displacement:

$$R_p = R\left(1 - \frac{x}{L}\right) \tag{4.92}$$

where $L$ is the length of an imaginary barrel in which the pressure force equals zero at its end. In reality, $L > L_b$, where $L_b$ is the length of the barrel. Referring to equation (1.6) and Figure 4.4, we can compose the following differential equation of motion for the projectile:

$$m\frac{d^2x}{dt^2} + C\frac{dx}{dt} + F = R(1 - \frac{x}{L}) \tag{4.93}$$

The initial conditions are:

$$\text{for} \quad t = 0 \quad x = 0, \quad \frac{dx}{dt} = 0 \qquad \text{(4.94)}$$

Applying conventional algebra, we restructure equation (4.93) into the following shape:

$$\frac{d^2x}{dt^2} + C\frac{dx}{dt} + \frac{R}{L}x + F = R \qquad \text{(4.95)}$$

Dividing equation (4.95) by $m$, we have:

$$\frac{d^2x}{dt^2} + 2n\frac{dx}{dt} + \omega_2^2 x + f = r \qquad \text{(4.96)}$$

where

$$\omega_2^2 = \frac{R}{Lm} \qquad \text{(4.96a)}$$

The Laplace domain of equation (4.96) at the initial conditions (4.94) reads:

$$l^2 x(l) + 2nl x(l) + \omega_2^2 x(l) = r - f \qquad \text{(4.97)}$$

The solution of equation (4.97) in the Laplace domain is:

$$x(l) = \frac{r - f}{l^2 + 2nl + \omega_2^2} \qquad \text{(4.98)}$$

Using pair 13 from Table 3.1, we invert equation (4.98) into the time domain:

$$x = \frac{r - f}{\omega_2^2}[1 - e^{-nt}(\cos\omega_3 t + \frac{n}{\omega_3}\sin\omega_3 t)] \qquad \text{(4.99)}$$

where

$$\omega_3^2 = \omega_2^2 - n^2 \qquad \text{(4.100)}$$

We assume that $\omega_3^2 > 0$. The cases when $\omega_2^2 - n^2 \leq 0$ were considered earlier for a similar situation.

Differentiating equation (4.100), we obtain the velocity:

$$\frac{dx}{dt} = \frac{r - f}{\omega_3} e^{-nt} \sin \omega_3 t \qquad (4.101)$$

Taking the first derivative from equation (4.101), we determine the acceleration:

$$\frac{d^2 x}{dt^2} = \frac{r - f}{\omega_3} e^{-nt} (\omega_3 \cos \omega_3 t - n \sin \omega_3 t) \qquad (4.102)$$

Equations (4.93) and, consequently, (4.99), (4.101), and (4.102) are applicable to the moment when the displacement becomes equal to the length of the barrel. At this point, the projectile reaches its maximum velocity and leaves the barrel. The projectile then continues to move due to its kinetic energy.

In order to analyze the parameters of motion, we can equate the left side of equation (4.99) to $L_b$ and try to determine the time the projectile moves inside of the barrel. If the time is determined, we can substitute it into equation (4.101) and calculate the projectile's maximum velocity. We also can estimate the role of the parameters involved, and finally make an appropriate decision regarding the parameters and performance of the system. However, equation (4.99) is highly transcendental and currently cannot be solved in general terms. Transcendental equations can be solved by graphical or numerical methods. Both methods are based on quantitative data that is not available at the preliminary stages of the analysis. The entire purpose of the simplified approach is to understand the interaction between the parameters and to determine the approximate outcomes during the early stages of the investigation.

In cases when highly transcendental equations obstruct the continuation of the simplified approach, we should look closely at the factors that cause their transcendence, with the goal of eliminating the less influential factors. In our case, the damping and stiffness forces are the relevant factors. The friction force has a constant value; it acts all the time during the projectile's motion in the barrel. Numerous studies show that the friction force causes some wear of the barrel, which should be replaced after a certain number of shoots.

Thus, the friction forces are significant. Air resistance at the beginning of the projectile's motion equals zero and increases proportionally to the increase of the velocity. Also, considering the relatively small cross-sectional area of the projectile, we accept that the damping force is much less influential than the friction force. So, it is reasonable to eliminate the damping force from the analysis by equating the damping factor $n$ to zero in equation (4.99). Based on this and equation (4.100), we obtain the modified expression for the displacement:

$$x = \frac{r-f}{\omega_2^2}(1 - \cos\omega_2 t) \qquad (4.103)$$

The modified equation for the velocity reads:

$$\frac{dx}{dt} = \frac{r-f}{\omega_2}\sin\omega_2 t \qquad (4.104)$$

Equating the left side of equation (4.103) to $L_b$ and denoting the traveling time during the process of acceleration as $t_3$, we have:

$$1 - \frac{L_b\omega_2^2}{r-f} = \cos\omega_2 t_3 \qquad (4.105)$$

Based on equation (4.105), we determine:

$$\sin\omega_2 t_3 = \frac{\omega_2}{r-f}\sqrt{L_b\left[2(r-f)-L_b\omega_2^2\right]} \qquad (4.106)$$

Combining equations (4.84) and (4.86), we calculate the projectile's maximum velocity:

$$v_{max} = \sqrt{L_b[2(r-f)-L_b\omega_2^2)}$$

Substituting the corresponding notations into this equation, we have:

$$v_{max} = \sqrt{\frac{L_b(R-2F)}{m}} \qquad (4.107)$$

It is interesting to compare the value of the maximum velocity according to equation (4.107) with the value of the maximum velocity $v_{1max}$ for a case where we assume the pressure in the barrel does not decrease during the process of acceleration. Applying energy considerations to this case, we write:

$$(R - F)L_b = \frac{m}{2} v_{1\,max}^2$$

From this expression, we obtain:

$$v_{1max} = \sqrt{\frac{2L_b(R - F)}{m}} \tag{4.108}$$

Realizing that the friction force exerted by the barrel is relatively small in comparison with the active force of the pressurized gases, we find that:

$$\frac{v_{1max}}{v_{max}} \cong \sqrt{2}$$

The actual maximum velocity of the projectile due to the damping resistance will be less than the value calculated according to equation (4.107). However, the results of this analysis let us make justifiable decisions regarding the parameters and the performance of the system.

### 4.5  Reciprocation Cycle of a Spring-Loaded Sliding Link

Numerous mechanical systems have mechanisms with reciprocating spring-loaded sliding links. In pneumatic and hydraulic machinery, these links represent spool valves and pistons; in some firearms they play the role of carriers; and in electromagnetic devices they are used in solenoids, etc. Despite these different functions and structural compositions, the cycles of motion among the sliding links are very similar. As a result, we can apply the same analytical description of their motion to most of them.

The cycle of a sliding spring-loaded reciprocating link consists of a forward and a backward stroke. Usually the link's forward stroke is caused by an active force, an impact, or an impulse. The active force could be generated by cyclically pressurized fluids, or by electro-magnetic factors. The impact cyclically occurs during the collision of the sliding link with another component. The impulse could be a result of an instantaneous injection of highly pressurized gases. During the forward stroke, the spring becomes deformed and accumulates potential energy. The backward stroke of the link is caused by the potential energy of the deformed spring. There is no external loading on the sliding link during the backward stroke.

The length of the stroke as well as the parameters of the sliding link and the spring are usually predetermined from design considerations. However, the time duration of the forward and backward stroke, the stresses in the spring, the required external loading factors, and the required initial velocity of the sliding link cannot be obtained from these considerations. All the information about these parameters can be obtained on the basis of the analytical approach to the link's cyclic motion.

We want to analyze the motion of the sliding link during its forward and backward strokes. Let's consider two options related to the forward stroke. In the first option, the motion of the link is caused by a constant active force whereas, in the second option, the link instantaneously obtains an initial velocity due to impact or impulse.

### 4.5.1 Forward Stroke Due to a Constant Force

During the forward stroke, the sliding link, which has a mass $m$, is subjected to the action of an active constant force $R$ and resisting forces of friction and stiffness as well. We assume that the air resistance is insignificant. The mass of the sliding link, the friction force $F$, and the stiffness coefficient $K$ are initially determined based on design considerations. Figure 4.5 represents the forward stroke when the horizontal motion of the sliding link 1, having a mass $m$ on a frictional surface, is caused by a constant force.

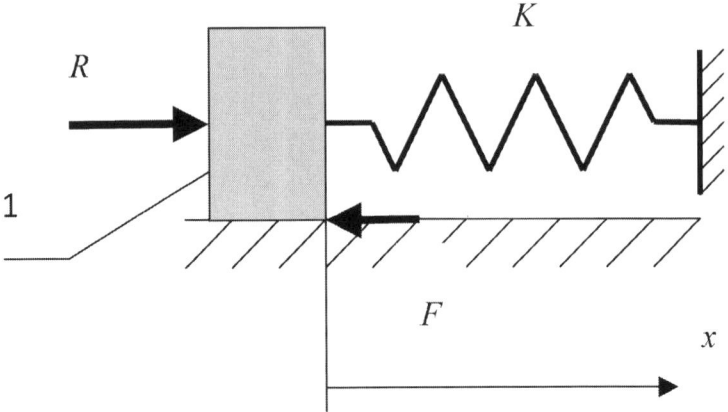

**Figure 4.5  The forward stroke is initiated by a constant force.**

Based on the model represented in Figure 4.5 and referring to equation (1.6), we can write the following differential equation of motion of the sliding link:

$$m\frac{d^2x}{dt^2} + Kx + F = R \tag{4.109}$$

The initial conditions are:

$$\text{for} \quad t = 0 \quad x = 0; \quad \frac{dx}{dt} = 0 \tag{4.110}$$

Dividing the expression (4.109) by $m$, we have:

$$\frac{d^2x}{dt^2} + \omega_0^2 x + f = r \tag{4.111}$$

The Laplace domain of equation (4.111) is:

$$l^2 x(l) + \omega_0^2 x(l) + f = r \tag{4.112}$$

The Laplace domain solution of equation (4.112) reads:

$$x(l) = \frac{r - f}{l^2 + \omega_0^2} \tag{4.113}$$

The inversion of equation (4.113) into the time domain is:

$$x = \frac{r-f}{\omega_0^2}\left(1 - \cos\omega_0 t\right) \tag{4.114}$$

Taking the first derivative from equation (4.114), we determine the link's velocity:

$$\frac{dx}{dt} = \frac{r-f}{\omega_0}\sin\omega_0 t \tag{4.115}$$

Equation (4.115) is a periodical function of time; hence, the link's velocity will increase to its maximum value and then decrease. At the end of the forward stroke, the sliding link's velocity equals zero. Setting the left side of equation (4.115) equal to 0, we have:

$$\sin\omega_0 t_r = 0 \tag{4.116}$$

where $t_r$ is the time duration of the forward stroke for this case. Because equation (4.116) is a periodical function, we may write:

$$\omega_0 t_r = i\pi \tag{4.117}$$

where

$$i = 0, 1, 2, 3, \ldots$$

The case when $i = 0$ corresponds to the initial conditions. Therefore, we have to take the next integer which is $i = 1$. Substituting this number into equation (4.117), we determine the time duration of the forward stroke:

$$t_r = \frac{\pi}{\omega_0} \tag{4.118}$$

As discussed earlier, the length of the forward stroke $S$ is predetermined. By the end of the forward stroke, the displacement of the link equals to $S$. Keeping this in mind and combining equations (4.114) and (4.118), we have:

$$S = \frac{r-f}{\omega_0^2}(1 - \cos\pi)$$

or

$$S = \frac{2(r-f)}{\omega_0^2} \qquad (4.119)$$

From equation (4.119), we determine the value of the required active force:

$$R = \frac{1}{2}KS + F \qquad (4.120)$$

According to equation (4.120), the force $R$ consists of a component related to the deformation of the spring and another associated with the friction. By the end of the forward stroke, the spring is deformed by a force $P_{smax}$ that equals:

$$P_{smax} = KS \qquad (4.121)$$

At the beginning of the link's displacement, the deformation process of the spring starts from zero loading and increases to $P_{smax}$; this is why this force is twice the deformation force in equation (4.120). The force $R$ acts all the time during the forward stroke. At the beginning of the forward stroke, this force accelerates the link to its maximum velocity. During the acceleration process, the link obtains a certain amount of kinetic energy. The link then decelerates, spending the accumulated kinetic energy while the force $R$ continues to act all the time to the end of the forward stroke.

The stress analysis of the spring should be based on the force according to equation (4.121).

### 4.5.2  Forward Stroke Due to Initial Velocity

Figure 4.6 illustrates the forward stroke for the case when the motion of the sliding link is initiated by a certain initial velocity $v_0$.

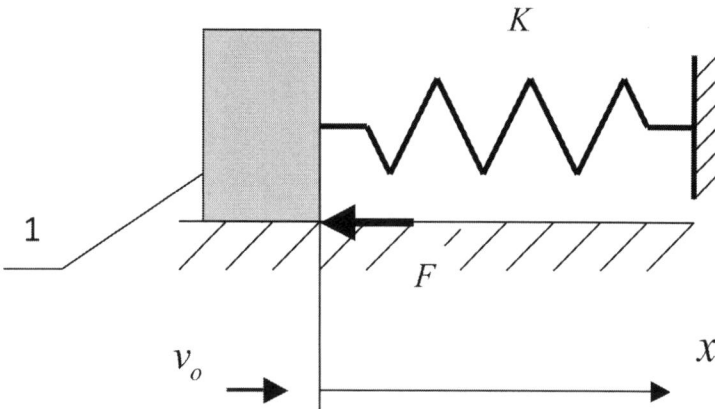

**Figure 4.6  The forward stroke is caused by the initial velocity of the sliding link.**

According to Figure 4.6, and referring to equation (1.1), we can compose the following differential equation of motion of the sliding link:

$$m\frac{d^2x}{dt^2} + Kx + F = 0 \qquad (4.122)$$

The initial conditions are:

$$\text{for} \quad t = 0 \quad x = 0; \quad \frac{dx}{dt} = v_0 \qquad (4.123)$$

Dividing equation (4.122) by $m$, we obtain:

$$\frac{d^2x}{dt^2} + \omega_0^2 x + f = 0 \qquad (4.124)$$

Converting equation (4.124) into the Laplace domain, we write:

$$l^2 x(l) - l v_0 + \omega_0^2 x(l) + f = 0 \qquad (4.125)$$

The solution of equation (4.125) in the Laplace domain is:

$$x(l) = \frac{l v_0}{l^2 + \omega_0^2} - \frac{f}{l^2 + \omega_0^2} \qquad (4.126)$$

Inverting equation (4.126) into the time domain, we obtain:

$$x = \frac{v_0}{\omega_0}\sin\omega_0 t - \frac{f}{\omega_0^2}(1-\cos\omega_0 t) \qquad (4.127)$$

Applying conventional algebra to equation (4.127), we write:

$$x = \frac{1}{\omega_0^2}(v_0\omega_0\sin\omega_0 t + f\cos\omega_0 t) - \frac{f}{\omega_0^2} \qquad (4.128)$$

Denoting:

$$\sin\beta_0 = \frac{f}{\sqrt{v_0^2\omega_0^2 + f^2}} \qquad (4.129)$$

and

$$\cos\beta_0 = \frac{v_0\omega_0}{\sqrt{v_0^2\omega_0^2 + f^2}} \qquad (4.130)$$

we transform equation (4.128) into a shape that is more suitable for analysis:

$$x = \frac{\sqrt{v_0^2\omega_0^2 + f^2}}{\omega_0^2}\sin(\omega_0 t + \beta_0) - \frac{f}{\omega_0^2} \qquad (4.131)$$

Differentiating equation (4.131), we determine the velocity of the link:

$$\frac{dx}{dt} = \frac{\sqrt{v_0^2\omega_0^2 + f^2}}{\omega_0}\cos(\omega_0 t + \beta_0) \qquad (4.132)$$

At the end of the forward stroke, the velocity equals zero. Equating the left side of equation (4.132) to zero, we obtain:

$$\cos(\omega_0 t_v + \beta_0) = 0 \qquad (4.133)$$

where $t_v$ is the time duration of the forward stroke for this case. From equation (4.133), we have:

$$\omega_0 t_v + \beta_0 = \frac{\pi}{2} \qquad (4.134)$$

Solving equation (4.134) for time $t_v$, we obtain:

$$t_v = \frac{0.5\pi - \beta_0}{\omega_0} \qquad (4.135)$$

Combining equation (4.131) with equation (4.135), and accounting that at the end of the forward stroke the displacement equals $S$, we have:

$$S = \frac{1}{\omega_0^2}(\sqrt{v_0^2\omega_0^2 + f^2} - f) \qquad (4.136)$$

From equation (4.136), we calculate the value of the required initial velocity $v_0$:

$$v_0 = \sqrt{\frac{S}{m}(SK + F)} \qquad (4.137)$$

Differentiating equation (4.132), we determine the acceleration (actually it is the deceleration):

$$\frac{d^2 x}{dt^2} = -\sqrt{v_0^2\omega_0^2 + f^2}\ \sin(\omega_0 t + \beta_0) \qquad (4.138)$$

According to equation (4.134), at the end of the forward stroke we have:

$$\sin(\omega_0 t_v + \beta_0) = 1 \qquad (4.139)$$

Combining equations (4.136), (4.137), (4.138), and (4.139), we determine the maximum value of the deceleration $a_{max}$ at the end of the forward stroke:

$$a_{max} = -\frac{1}{m}\sqrt{KS(KS + F) + F^2}$$

If we set the friction force equal to zero in this equation — it is not applied to the spring — we obtain the maximum value of the spring deformation force that is determined by equation (4.121).

### 4.5.3 Backward Stroke

At the end of the forward stroke, the link comes to a standstill point and the spring is compressed. The link starts to move into the opposite direction, causing the friction force to change direction. The velocity of the link becomes negative, and the friction force is directed positively. As a rule, the friction force is always directed into the opposite direction of the velocity. Figure 4.7 illustrates the backward stroke of the link. As it shows, at the beginning of the forward stroke the sliding link is located in position 1, at the origin 0 of the axis $x$, and the spring is not deformed. By the end of the forward stroke, the link is in position 2, the spring is deformed according to the displacement $S$, and the link will start to move in the negative direction.

Because the friction force at the beginning of the backward stroke instantaneously changes its direction, the motion of the link cannot be described by equation (4.132), which is derived for the

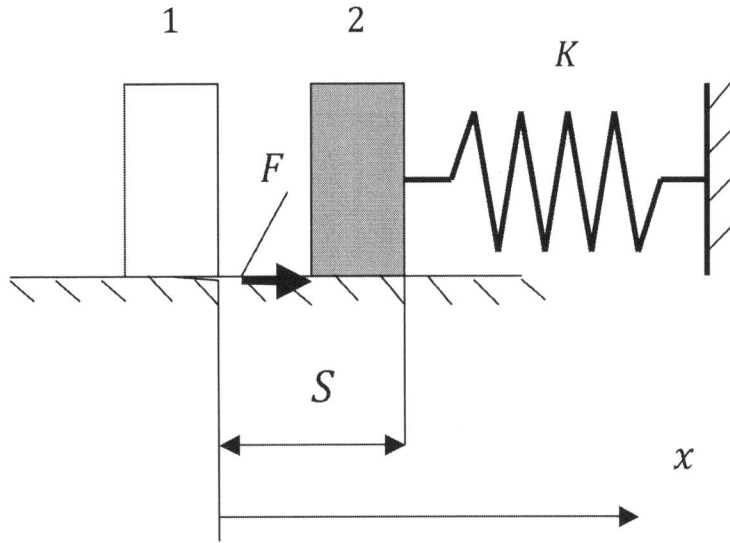

**Figure 4.7  The backward stroke of the sliding link.**

forward stroke. This equation cannot by itself change the direction of the friction force. We must compose a differential equation of motion appropriate for describing the backward stroke of the sliding link. Referring to equation (1.1) and Figure 4.7, we can compose the following differential equation of motion for this case:

$$m\frac{d^2x}{dt^2} + Kx + F = 0 \tag{4.140}$$

The initial conditions are:

$$\text{for} \quad t = 0 \quad x = S; \quad \frac{dx}{dt} = 0 \tag{4.141}$$

We can see that equations (4.123) and (4.140) are identical, but they describe different strokes that reflect different processes of motion. Because these two equations have different initial conditions of motion, they have different solutions. This illustrates the idea that the same differential equation may have many applications.

Omitting a few trivial steps, we write the Laplace domain for equation (4.140) at the initial conditions (4.141):

$$l^2x(l) - l^2S + \omega_0^2x(l) + f = 0 \tag{4.142}$$

The Laplace domain solution of equation (4.142) is:

$$x(l) = \frac{l^2S}{l^2 + \omega_0^2} - \frac{f}{l^2 + \omega_0^2} \tag{4.143}$$

The inversion of equation (4.143) into the time domain reads:

$$x = \frac{1}{\omega_0^2}[(S\omega_0^2 + f)\cos\omega_0 t - f)] \tag{4.144}$$

At the end of the backward stroke, the displacement equals zero. Setting the left side of equation (4.144) equal to zero, we determine the time duration $t_b$ of the backward stroke:

$$t_b = \frac{1}{\omega_0}\cos^{-1}(\frac{f}{S\omega_0^2 + f}) \tag{4.145}$$

Based on equations (4.119), (4.136), and (4.145), we determine the duration of the reciprocation cycle for the two versions considered above.

Basically this concludes our analysis of the backward stroke. The velocity and acceleration during this stroke do not influence the decisions regarding selection of the values for the system's parameters.

## 4.6  Pneumatically Operated Soil Penetrating Machines

Pneumatically operated, self-propelled, soil penetrating machines are widely used for underground horizontal borehole making. Figure 4.8 illustrates this machine. The machine has a tubular body 3 that contains a reciprocating striker 2 and an air distributing mechanism 1. The compressed air from the compressor enters the air distributing mechanism and creates a pressure force $R_v$, causing the motion of the striker 2 during its forward stroke. The air in front of the striker, being pushed through the duct 4 into the atmosphere, develops a damping resistance that is represented in Figure 4.8 by the dashpot having a damping coefficient C. A resisting friction force $F$ is applied to the striker from the inner surface of the tubular body. At the end of the forward stroke $S_v$, the striker imparts

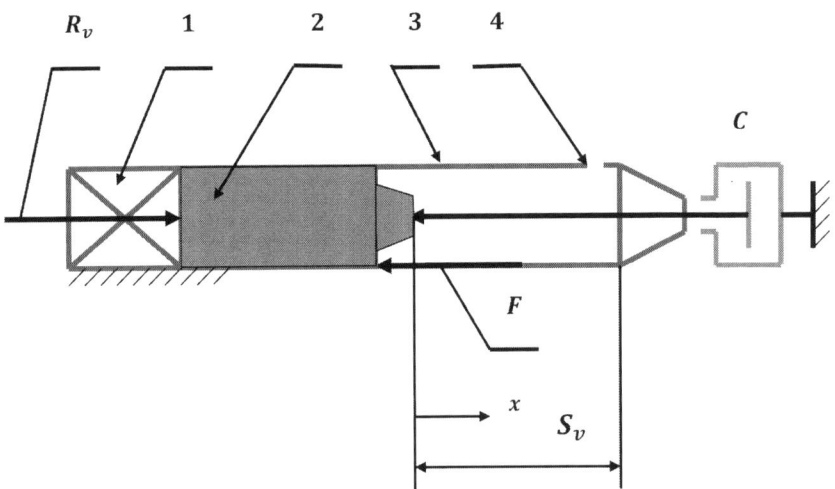

**Figure 4.8  Pneumatically operated, self-propelled, soil penetrating machine.**

a blow to the inner forehead of the tubular body, resulting in an incremental penetration of the body into the soil. Our goal is to determine the kinetic energy of the striker at the end of the forward stroke. We also want to calculate the time duration of this stroke. We assume that during the forward stroke the tubular body remains motionless.

In this type of pneumatically operated machines, the pressure in the rear chamber behind the accelerating striker is gradually dropping. This occurs because the small cross-sectional areas of the air ducts restrict the flow of the compressed air into the rear chamber behind the striker. The air flow through these ducts cannot keep up with the chamber's rapidly increasing volume during the striker's acceleration. As a result, the active force $R_v$ decreases proportionally as the velocity of the striker increases. Let's assume that in this case the active pressure force is a linear function of the striker's velocity. According to this assumption, we may write:

$$R_v = R - C_v \frac{dx}{dt} \qquad (4.146)$$

where $R$ is the maximum value of the pressure force and $C_v$ is the damping coefficient related to the striker's forward stroke.

Based on Figure 4.8 and equation (1.6), we may write the following differential equation of horizontal motion of the striker:

$$m \frac{d^2x}{dt^2} + C \frac{dx}{dt} + F = R - C_v \frac{dx}{dt} \qquad (4.147)$$

where $m$ is the mass of the striker, and the rest of the parameters are those introduced above. Dividing equation (4.147) by $m$, we have:

$$\frac{d^2x}{dt^2} + 2n \frac{dx}{dt} + f = 1 - 2n_1 \frac{dx}{dt} \qquad (4.148)$$

where

$$2n_1 = \frac{C_v}{m} \qquad (4.149)$$

Combining equations (4.148) and (4.149), we may write:

$$\frac{d^2x}{dt^2} + 2n_0\frac{dx}{dt} + f = r \qquad (4.150)$$

where

$$n_0 = n + n_1 \qquad (4.151)$$

The initial conditions of motion for equation (4.147) are:

$$\text{for} \quad t = 0 \quad x = 0; \quad \frac{dx}{dt} = 0 \qquad (4.152)$$

The Laplace domain expression for equation (4.150) at the initial conditions (4.152) reads:

$$l^2x(l) + 2n_0lx(l) + f = r \qquad (4.153)$$

The Laplace domain solution of equation (4.153) is:

$$x(l) = \frac{r - f}{l(l + 2n_0)} \qquad (4.154)$$

The right side of equation (4.154) does not have an equivalent representation in Table 3.1. However, referring to equation (3.63), we may rewrite equation (4.154) in the way that allows using this table:

$$x(l) = \frac{r - f}{2n_0l} - \frac{r - f}{2n_0[l - (2n_0)]} \qquad (4.155)$$

The inversion of equation (4.155) into the time domain reads:

$$x = \frac{(r - f)t}{2n_0} + \frac{r - f}{4n_0^2}\left(e^{-2n_0t} - 1\right) \qquad (4.156)$$

Taking the first derivative from equation (4.156), we obtain the expression for the velocity of the striker:

$$\frac{dx}{dt} = \frac{r - f}{2n_0}(2 - e^{-2n_0t}) \qquad (4.157)$$

According to Figure 4.8, the displacement of the striker at the end of the forward stroke equals $S_v$. Setting the left side of equation (4.156) equal to $S_v$, we can determine the duration of time $T$ of the forward stroke:

$$S_v = \frac{(r-f)T}{2n_0} + \frac{r-f}{4n_0^2}\left(e^{-2n_0T} - 1\right) \qquad (4.158)$$

This expression is transcendental; it cannot be solved with respect to $T$. In order to get an approximate solution, it is possible to represent the exponential function $e^{-2n_0T}$ as a Taylor series:

$$e^{-2n_0T} = 1 - 2n_0T + \frac{(2n_0T)^2}{2} - \frac{(2n_0T)^3}{6} + \ldots \qquad (4.159)$$

To simplify the solution, we incorporate in equation (4.158) just the first three terms of equation (4.159). After corresponding transformations, we determine the time duration of the forward stroke of the striker:

$$T = \sqrt{\frac{2s_v}{r-f}} \qquad (4.160)$$

Combining equations (4.157) and (4.160), we determine the velocity $v$ of the striker at the end of its forward stroke:

$$v = \frac{r-f}{2n_0}\left(2 - e^{-2n_0\sqrt{\frac{2S_v}{r-f}}}\right) \qquad (4.161)$$

The kinetic energy of the striker at the end of the forward stroke equals:

$$E = \frac{mv^2}{2} \qquad (4.162)$$

Finally, using equations (4.161) and (4.162), we determine the impact energy $E$ of the striker:

$$E = \frac{m}{2}\left[\frac{r-f}{2n_0}\left(2 - e^{-2n_0\sqrt{\frac{2S_v}{r\,f}}}\right)\right]^2$$

This concludes our analysis of pneumatically operated, self-propelled, soil penetration machines.

# PIECE-WISE LINEAR APPROXIMATION

There are currently no consistent methodologies for solving non-linear differential equations in general terms. The structures of linear and non-linear differential equations of motion are identical. In the majority of cases, the non-linearity is caused by the nature of the resisting forces. Basically, these resisting forces comprise the forces of inertia, damping, stiffness, and friction. In linear equations, the coefficients for acceleration, velocity, and displacement have constant values; if these coefficients are variables, then the equations become non-linear. If the displacement or its derivatives in the differential equation have a power higher than one, the equation is also non-linear. Because this type of equation is not typical for describing the motion of actual mechanical systems, it is not considered in this book. The problems with non-linear forces of inertia belong to a very specific group of mechanical systems and are also not included in this text. The friction force is almost always considered as a constant force. Thus, we will focus on non-linear damping and stiffness resisting forces.

In many cases, the exact analytical expression of a non-linear resisting force is unknown. Graphing the data about the resisting forces helps us understand their characteristics and estimate the degree of their non-linearity. Using selected methodologies, we can calculate the values of the resisting force at each point of the graph's curve, but in the majority of cases, these empiric expressions cannot be used with differential equations.

However, in many practical situations, piece-wise linear approximation allows us to overcome the problems of non-linearity and find adequate solutions. Essentially, we replace the experimental curve of the graph with a broken line. We can then analyze the motion by a series of successive linear differential equations corresponding to the broken line segments. On the first segment, the initial conditions of the motion are determined by the problem's formulation. The initial conditions of motion for the next segment are then set by the conditions of motion at the end of the previous segment.

Piece-wise linear approximation can be used even in cases when more than one resisting force is non-linear. However, the procedure of calculating the solution and analyzing it then becomes very cumbersome. In the majority of practical cases associated with dynamics of mechanical systems, just one non-linear resisting force is present; and it could be the damping resisting force or the stiffness resisting force. The damping force is exerted when the movable body interacts with fluids, whereas the stiffness force occurs during the interaction with an elastic medium.

In case of an interaction with a visco-elastic medium, these two resisting forces could be non-linear simultaneously. The acceleration/deceleration of an aircraft is also associated with two non-linear resisting forces, namely, the force of inertia and the damping force. It is perhaps possible to imagine a case when all three resisting forces (inertia, damping, and stiffness) are non-linear. Putting aside these exceptional cases, we will concentrate our attention on the majority of actual problems that are associated with damping or stiffness non-linearity.

## 5.1  Penetrating into an Elasto-Plastic Medium

In this section, we will analyze the interaction that takes place when a rigid body penetrates elasto-plastic materials such as steel, concrete, polymer, soil, and other media. For example, let's analyze the dynamic process associated when a projectile penetrates a vertical wall made of an elasto-plastic material. Assume the projectile's mass, the shape, and dimensions are predetermined and the characteristics of the material and thickness of the wall are known. Supposed the projectile has the shape of a sharpened cylinder. Our analysis should determine the initial velocity of the projectile at the beginning of its penetration and the time duration of this process. Assume that, after the projectile has completed its penetration through the wall, its velocity equals zero.

In this case, the resisting forces include a frontal resisting force applied to the projectile's forehead and a lateral frictional resisting force applied to the projectile's cylindrical surface. The friction force is caused by the elastic properties of the material; it has a constant value and is proportional to the projectile's diameter and length. The frontal resisting force represents the reaction to deformation (penetration); it is exerted by the material, according to the cross-sectional area of the projectile.

The stiffness force of an elasto-plastic material is often non-linear. Before we can apply piece-wise linear approximation, we must clarify the characteristics of this stiffness force. Fortunately, at this phase of the analysis there is no need to involve the quantitative values of the parameters. However, we must get the appropriate data regarding the frontal resistance forces exerted by this material during its penetration.

The data can usually be represented in the form of graphs that demonstrate the relationship between the frontal resistance forces and the material's deformation. The curved graph OABCD in Figure 5.1 represents the dependence of the frontal resistance force $P$ on the displacement $x$ of the penetrator. Actually the penetration is one form of deformation. The graph in Figure 5.1 is non-linear; we will use piece-wise linear approximate to solve the problem.

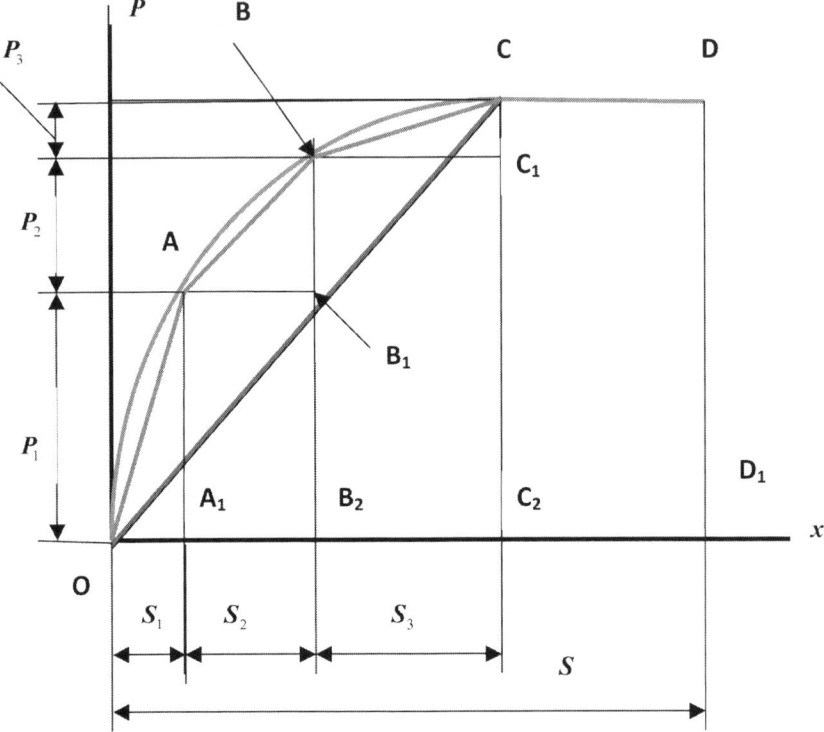

**Figure 5.1  The deformation (rheological) model
of an elasto-plastic material.**

The simplest way to use this methodology is to replace curve OABCD by a broken line OCD. We can then consider two intervals of motion. The differential equation of motion on the first interval is valid to the point where the projectile's displacement equals the distance from the origin O to the end of the segment $S_3$. During this interval, the projectile is subjected to a linear stiffness resisting force in accordance with line OC. According to Figure 5.1, the stiffness coefficient $K^*$ equals:

$$K^* = \frac{P_1 + P_2 + P_3}{S_1 + S_2 + S_3} \tag{5.1}$$

The initial conditions of motion for the first interval are based on the original formulation of the problem, which indicated a certain velocity for the projectile. The second interval for this case is characterized by a constant resisting force. As seen in Figure 5.1, this resisting force equals the sum of the following forces: $P_1 + P_2 + P_3$. The initial displacement on the second interval equals $S - S_1 - S_2 - S_3$, where S is the thickness of the wall. The initial velocity of motion for this second interval equals the ending velocity of the first interval.

Based on all these considerations, we can compose a linear differential equation of motion for each of these two intervals. In solving these equations, we perform the analysis needed to determine the required values of the parameters. Realize that replacing graph OABCD by broken line OCD will contribute an essential inaccuracy to the results. However, in real cases, it is justifiable to apply this two-interval piece-wise linear approximation in order to estimate the process's parameters. In many practical situations, this two-interval solution provides sufficient information for making engineering decisions.

Figure 5.1 shows that the amount of energy characterized by the area contained between curve OABC and straight line OC is not taken into consideration by the two-interval approach. Therefore, the required initial velocity that we calculate for the projectile is essentially less than it should be. Replacing the curve by a broken line with smaller segments will result in a more accurate solution. When we analyze a real-life problem, the lengths of the segments should be justified by appropriate considerations. For our purposes, we replace the curve OABC by a broken line OABC.

This approach increases the accuracy in comparison to the two-interval approach. Thus, we will perform a four-interval piece-wise linear approximation.

### 5.1.1  First Interval

In order to characterize the stiffness resisting force acting on the first interval, we must determine the stiffness coefficient for this interval. According to Figure 5.1, at the end of the first interval, the projectile's displacement equals $S_1$. Using triangle $OAA_1$ we determine the stiffness coefficient $K_1$ for this interval:

$$K_1 = \frac{P_1}{S_1} \tag{5.2}$$

Denoting by $x_1$ the displacement of the projectile on the first interval, we write:

$$0 \le x_1 \le S_1 \tag{5.3}$$

Based on Figure 5.1 and equation (1.1), we can compose the following differential equation of motion for the first interval:

$$m\frac{d^2x_1}{dt_1^2} + K_1x_1 + F = 0 \tag{5.4}$$

where $t_1$ is the running time of the first interval.

The initial conditions of motion for the first interval are:

$$\text{for} \quad t_1 = 0 \quad x_1 = 0; \quad \frac{dx_1}{dt_1} = v_0 \tag{5.5}$$

Dividing equation (5.4) by $m$, we have:

$$\frac{d^2x_1}{dt_1^2} + \omega_1^2 x_1 + f = 0 \tag{5.6}$$

where

$$\omega_1^2 = \frac{K_1}{m} \tag{5.7}$$

The Laplace domain of equation (5.6) at the initial conditions (5.5) reads:

$$l^2 x_1(l) - l v_0 + \omega_1^2 x_1(l) + f = 0 \tag{5.8}$$

The solution of equation (5.8) in the Laplace domain is:

$$x_1(l) = \frac{lv_0}{l^2 + \omega_1^2} - \frac{f}{l^2 + \omega_1^2} \tag{5.9}$$

Inverting equation (5.9) into the time domain, we have:

$$x_1 = \frac{v_0}{\omega_1}\sin\omega_1 t_1 - \frac{f}{\omega_1^2}(1 - \cos\omega_1 t_1) \tag{5.10}$$

We then transform equation (5.10) into the shape that is more suitable for analysis:

$$x_1 = \frac{1}{\omega_1^2}[\sqrt{v_0^2\omega_1^2 + f^2}\,\sin(\omega_1 t_1 + \gamma_1) - f] \tag{5.11}$$

where

$$\sin\gamma_1 = \frac{f}{\sqrt{v_0^2\omega_1^2 + f^2}} \tag{5.12}$$

and

$$\cos\gamma_1 = \frac{v_0\omega_1}{\sqrt{v_0^2\omega_1^2 + f^2}} \tag{5.13}$$

Differentiating equation (5.11), we determine the velocity on the first interval:

$$\frac{dx_1}{dt_1} = \frac{1}{\omega_1}\sqrt{v_0^2\omega_1^2 + f^2}\,\cos(\omega_1 t_1 + \gamma_1) \tag{5.14}$$

At the end of the first interval, the displacement equals $S_1$. Equating the left side of equation (5.10) to $S_1$, we determine the time $T_1$ of the duration of motion over the first interval:

$$T_1 = \frac{1}{\omega_1}[\sin^{-1}\left(\frac{S_1\omega_1 + f}{\sqrt{v_0^2\omega_1^2 + f^2}}\right) - \gamma_1] \tag{5.15}$$

Substituting the time according to equation (5.15) into equation (5.14), we determine the velocity $v_1$ at the end of the first interval:

$$v_1 = \frac{1}{\omega_1}\sqrt{v_0^2\omega_1^2 + f^2}\,\cos(\omega_1 T_1 + \gamma_1) \tag{5.16}$$

For this particular case, there is no need to determine the acceleration. According to equation (5.16), the displacement $S_1$ and the velocity $v_1$ represent the initial conditions of motion for the second interval.

### 5.1.2  Second Interval

As Figure 5.1 illustrates, the displacement on the second interval equals $S_2$. Therefore, the stiffness coefficient $K_2$ for the second interval is determined according to triangle $ABB_1$, and is:

$$K_2 = \frac{P_2}{S_2} \tag{5.17}$$

The kinetic energy of the projectile during the penetration process is spent on deforming the material. According to Figure 5.1, for the second interval of motion, the magnitude of the deformation energy is proportional to the area of trapezoid $A_1ABB_1B_2$. The deformation energy associated with stiffness resistance is proportional to the area of triangle $ABB_1$, which represents just a part of this trapezoid. The other part of this trapezoid is rectangle $A_1AB_1B_2$. According to this rectangle, the deformation energy is the product of multiplying the force $P_1$ by the displacement $S_2$. Therefore, the differential equation of motion on the second interval should include the constant resisting force $P_1$ along with the stiffness and friction resisting forces.

Denoting the displacement and time on the second interval by $x_2$ and $t_2$ respectively, and accounting for Figure 5.1 and equation (1.1), we can compose the following differential equation of motion:

$$m\frac{d^2x_2}{dt_2^2} + K_2x_2 + P_1 + F = 0 \tag{5.18}$$

The initial conditions are:

$$\text{for} \quad t_2 = 0 \quad x_2 = S_1; \quad \frac{dx_2}{dt_2} = v_1 \qquad (5.19)$$

Dividing equation (5.18) by $m$, we obtain:

$$\frac{d^2x_2}{dt_2^2} + \omega_2^2 x_2 + p_1 + f = 0, \qquad (5.20)$$

where

$$\omega_2^2 = \frac{K_2}{m} \qquad (5.21)$$

and

$$p_1 = \frac{P_1}{m} \qquad (5.22)$$

The Laplace domain of the equation (5.20) reads:

$$l^2 x_2(l) - lv_1 - l^2 S_1 + \omega_2^2 x_2(l) + p_1 + f = 0 \qquad (5.23)$$

The solution of equation (5.23) in the Laplace domain is:

$$x_2(p) = \frac{lv_1}{l^2 + \omega_2^2} + \frac{l^2 S_1}{l^2 + \omega_2^2} - \frac{p_1 + f}{l^2 + \omega_2^2} \qquad (5.24)$$

The inversion of equation (5.24) into the time domain reads:

$$x_2 = \frac{v_1}{\omega_2}\sin\omega_2 t_2 + S_1 \cos\omega_2 t_2 - \frac{p_1 + f}{\omega_2^2}(1 - \cos\omega_2 t_2) \qquad (5.25)$$

After performing the necessary transformations with equation (5.25), we obtain:

$$x_2 = \frac{1}{\omega_2^2}[\sqrt{v_1^2\omega_2^2 + (S_1\omega_2^2 + p_1 + f)^2}$$
$$\times \sin(\omega_2 t_2 + \gamma_2) - (p_1 + f)] \qquad (5.26)$$

where

$$\sin \gamma_2 = \frac{S_1 \omega_2^2 + p_1 + f}{\sqrt{v_1^2 \omega_2^2 + (S_1 \omega_2^2 + p_1 + f)^2}} \tag{5.27}$$

and

$$\cos \gamma_2 = \frac{v_1 \omega_2}{\sqrt{v_1^2 \omega_2^2 + (S_1 \omega_2^2 + p_1 + f)^2}} \tag{5.28}$$

Taking the first derivative from equation (5.26), we determine the velocity on the second interval:

$$\frac{dx_2}{dt_2} = \frac{1}{\omega_2} \sqrt{v_1^2 \omega_2^2 + (S_1 \omega_2^2 + p_1 + f)^2} \, \cos(\omega_2 t_2 + \gamma_2) \tag{5.29}$$

At the end of the second interval, the displacement equals $S_1 + S_2$. Equating the left side of equation (5.26) to this sum, we calculate the time duration $T_2$ of motion on the second interval:

$$T_2 = \frac{1}{\omega_2} \left\{ \sin^{-1} \left[ \frac{(S_1 + S_2) \omega_2^2 + p_1 + f}{\sqrt{v_1^2 \omega_2^2 + (S_1 \omega_2^2 + p_1 + f)^2}} \right] - \gamma_2 \right\} \tag{5.30}$$

Substituting the time according to equation (5.30) into equation (5.29), we determine the velocity $v_2$ at the end of the second interval:

$$v_2 = \frac{1}{\omega_2} \sqrt{v_1^2 \omega_2^2 + (S_1 \omega_2^2 + p_1 + f)^2} \, \cos(\omega_2 T_2 + \gamma_2) \tag{5.31}$$

The velocity according to equation (5.31) is the initial velocity of the motion for the third interval.

### 5.1.3  Third Interval

According to triangle $BC_1C$ in Figure 5.1, we determine the stiffness coefficient $K_3$ for the third interval:

$$K_3 = \frac{P_3}{S_3} \tag{5.32}$$

The deformation energy on this interval is proportional to the area of the trapezoid $B_2B_1BC_1C_2$. The stiffness coefficient from equation (5.32) is associated with the deformation energy that is related to the triangle $BCC_1$. At the same time, the constant force representing the sum of forces $P_1 + P_2$ is related to the rectangular area $B_2B_1BCC_1C_2$ and is associated with the energy deformation of this rectangle. Consequently, the resisting force $P_1 + P_2$ should be included in the differential equation of motion on the third interval. Accounting for all these considerations and referring to equation (1.1) — and also denoting the displacement and time for the third interval by $x_3$ and $t_3$ respectively, we can compose the following differential equation of motion for this interval:

$$m\frac{d^2x_3}{dt_3^2} + K_3x_3 + P_1 + P_2 + F = 0 \qquad (5.33)$$

The initial conditions are:

$$\text{for} \quad t_3 = 0 \quad x_3 = S_1 + S_2; \quad \frac{dx_3}{dt_3} = v_2 \qquad (5.34)$$

Because equation (5.33) and its initial conditions (5.34) are similar to equation (5.18) and its initial conditions (5.19) for the second interval, we omit the intermediate mathematical procedures and, based on the equation (5.26), we may write:

$$x_3 = \frac{1}{\omega_3^2}[\sqrt{v_2^2\omega_3^2 + (S_1\omega_3^2 + S_2\omega_3^2 + p_1 + p_2 + f)^2}$$
$$\times \sin(\omega_3t_3 + \gamma_3) - (p_1 + p_2 + f)] \qquad (5.35)$$

where

$$\omega_3^2 = \frac{K_3}{m}; \qquad (5.36)$$

$$p_2 = \frac{P_2}{m}; \qquad (5.37)$$

$$\sin \gamma_3 = \frac{S_1\omega_3^2 + S_2\omega_3^2 + p_1 + p_2 + f}{\sqrt{v_2^2\omega_3^2 + (S_1\omega_3^2 + S_2\omega_3^2 + p_1 + p_2 + f)^2}} \; ; \qquad (5.38)$$

$$\cos \gamma_3 = \frac{v_2\omega_3}{\sqrt{v_2^2\omega_3^2 + (S_1\omega_3^2 + S_2\omega_3^2 + p_1 + p_2 + f)^2}} \qquad (5.39)$$

Differentiating equation (5.35), we obtain the velocity for the third interval:

$$\frac{dx_3}{dt_3} = \frac{1}{\omega_3} \sqrt{v_2^2\omega_3^2 + (S_1\omega_3^2 + S_2\omega_3^2 + p_1 + p_2 + f)^2} \qquad (5.40)$$
$$\times \cos(\omega_3 t_3 + \gamma_3)$$

At the end of this interval, the displacement of the projectile equals the following sum of displacements: $S_1 + S_2 + S_3$. Setting the left side of equation (5.35) equal to this sum, we calculate the duration of time $T_3$ of motion for the third interval:

$$T_3 = \frac{1}{\omega_3} \left\{ \sin^{-1} \left[ \frac{(S_1 + S_2 + S_3)\omega_3^2 + p_1 + p_2 + f)}{\sqrt{v_2^2\omega_3^2 + (S_1\omega_3^2 + S_2\omega_3^2 + p_1 + p_2 + f)^2}} \right] \quad (5.41) \right.$$
$$\left. - \gamma_3 \right\}$$

Substituting the time according to equation (5.41) into equation (5.40), we determine the velocity $v_3$ at the end of the third interval:

$$v_3 = \frac{1}{\omega_3} \sqrt{v_2^2\omega_3^2 + (S_1\omega_3^2 + S_2\omega_3^2 + p_1 + p_2 + f)^2} \qquad (5.42)$$
$$\times \cos(\omega_3 T_3 + \gamma_3)$$

The velocity according to equation (5.42) represents the initial velocity for the fourth (last) interval.

### 5.1.4  Fourth Interval

This interval is characterized by development of plastic deformations at a constant resisting force which, according to Figure 5.1, equals the sum of the following forces: $P_1 + P_2 + P_3$. Denoting the displacement and time for this interval as $x_4$ and $t_4$ respectively and referring to equation (1.1), we can compose the following differential equation of motion for this interval:

$$m\frac{d^2 x_4}{dt_4^2} + P_1 + P_2 + P_3 + F = 0 \qquad (5.43)$$

The initial conditions are:

$$\text{for} \quad t_4 = 0; \quad x_4 = S_1 + S_2 + S_3; \quad \frac{dx_4}{dt_4} = v_3 \qquad (5.44)$$

Dividing the equation (5.44) by $m$, we have:

$$\frac{d^2 x_4}{dt_4^2} + p_1 + p_2 + p_3 + f = 0 \qquad (5.45)$$

where

$$p_3 = \frac{P_3}{m} \qquad (5.46)$$

The Laplace domain of equation (5.45) at the initial conditions (5.44) is:

$$l^2 x_4(l) - v_3 l - l^2\left(S_1 + S_2 + S_3\right) + p_1 + p_2 + p_3 + f = 0 \qquad (5.47)$$

The Laplace domain solution of equation (5.47) reads:

$$x_4(l) = S_1 + S_2 + S_3 + \frac{v_3}{l} - \frac{p_1 + p_2 + p_3 + f}{l^2} \qquad (5.48)$$

The inversion of equation (5.48) into the time domain is:

$$x_4 = S_1 + S_2 + S_3 + v_3 t_4 - \frac{1}{2}(p_1 + p_2 + p_3 + f)t_4^2 \qquad (5.49)$$

Taking the first derivative from equation (5.49), we determine the velocity on this interval:

$$\frac{dx_4}{dt_4} = v_3 - (p_1 + p_2 + p_3 + f)t_4 \qquad (5.50)$$

At the end of the fourth interval, the projectile's velocity may equal zero. Thus, setting the left side of equation (5.50) equal to zero, we determine the time duration $T_4$ of motion on this interval:

$$T_4 = \frac{v_3}{p_1 + p_2 + p_3 + f} \tag{5.51}$$

The projectile's displacement at the end of the fourth interval equals the thickness S of the wall. Setting the left side of equation (5.49) equal to S and substituting into this equation the time according to equation (4.51), we determine the value of the velocity $v_3$:

$$v_3 = \sqrt{2(S - S_1 - S_2 - S_3)(p_1 + p_2 + p_3 + f)} \tag{5.52}$$

In order to determine the required value of the velocity $v_0$, it is necessary to go in reverse order through the analysis of motion on the intervals. We start by setting the left side of equation (5.42) (which represents the velocity at the end of the third interval) equal to the obtained value according to equation (5.52). This step will result in a transformed expression for the velocity at the end of the second interval; it should then be combined with equation (5.31). Finally, combining the last result with equation (5.16) from the first interval, we can determine the required value of velocity $v_0$. The actual value of the projectile's initial velocity should exceed the calculated value.

The time duration of the penetration process represents the sum of the particular durations of time for each interval.

Note that the required value of the projectile's initial velocity $v_0$ can be calculated in a much shorter way. The energy that is required to penetrate the wall consists of the sum of the energy deformation $E_D$ and energy to overcome the friction $E_F$. The deformation energy can be calculated according to the area contained between the broken line $OABCDD_1$ and axis $x$ in Figure 5.1. Thus, we can write:

$$\begin{aligned}
E_D &= \frac{1}{2}P_1S_1 + \frac{1}{2}P_2S_2 + P_1S_2 + \frac{1}{2}P_3S_3 + (P_1 + P_2)S_3 \\
&+ (P_1 + P_2 + P_3)(S - S_1 - S_2 - S_3)
\end{aligned} \tag{5.53a}$$

The energy spent to overcome the friction resistance $E_f$ in the process of penetration is:

$$E_F = FS \qquad (5.53b)$$

We then calculate the total kinetic energy $E$ that the projectile should possess at the beginning of the penetration process:

$$E = E_D + E_F \qquad (5.53c)$$

However, the total kinetic energy of the projectile also equals:

$$E = \frac{1}{2} mv_0^2 \qquad (5.53)$$

Combining equations (5.53a), (5.53b), and (5.53c) with equation (5.53), we obtain:

$$\frac{1}{2} mv_0^2 = \frac{1}{2} P_1 S_1 + \frac{1}{2} P_2 S_2 + P_1 S_2 + \frac{1}{2} P_3 S_3 + (P_1 + P_2) S_3$$
$$+ (P_1 + P_2 + P_3)(S - S_1 - S_2 - S_3) + FS \qquad (5.54)$$

Applying the appropriate algebraic procedures to equation (5.54), we obtain the expression for calculating the value of $v_0$

We must also determine the duration of the time for the penetration process. Therefore, it becomes necessary to analyze the motion on the intervals presented above. However, our main purpose here was to demonstrate piece-wise linear approximation for a case with a non-linear stiffness resisting force.

## 5.2  Non-Linear Damping Resistance

In an earlier discussion, the damping force applied to a ship was linear and combined water and air resistance. However, in real conditions, this force is not linear. Appropriate data presented in the form of graphs demonstrate the dependence of the combined damping

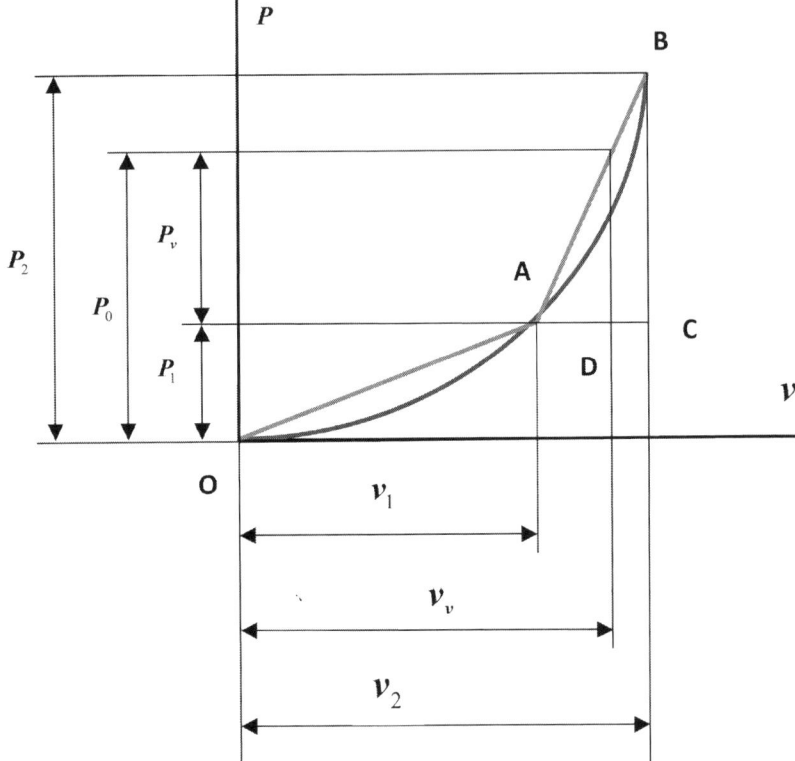

**Figure 5.2  The graph of a non-linear damping resisting force.**

force on the ship's velocity, as shown in Figure 5.2 where curve OAB represents the relationship between damping force $P$ and velocity $v$. Let's assume that this graph represents the actual damping resistance to a vessel's motion.

The graph should not be applied for velocities higher than $v_2$ and, consequently, the maximum velocity of the ship should not exceed $v_2$.

Replacing the curved graph by the broken line OAB, we can use piece-wise linear approximation to analyze the dynamics of the vessel, considering its motion on two intervals. The problem in this case is to calculate the power needed to accelerate the vessel to velocity $v_v$ while $v_1 < v_v \leq v_2$, where $v_1$ and $v_2$ are the velocities at the end of the first and second intervals respectively.

## 5.2.1  First Interval

In accordance with Figure 5.2, we determine the damping coefficient $C_1$ for the first interval using the following expression:

$$C_1 = \frac{P_1}{v_1} \qquad (5.55)$$

where $P_1$ is the maximum damping force at the end of the first interval.

Assuming that the vessel is accelerated by a constant active force and referring to both equation (1.6) and Figure 5.2, we can compose the following differential equation of motion for the first interval:

$$m\frac{d^2x_1}{dt_1^2} + C_1\frac{dx_1}{dt_1} = R \qquad (5.56)$$

The initial conditions for this interval are:

$$\text{for} \quad t_1 = 0; \quad x_1 = 0; \quad \frac{dx_1}{dt_1} = 0 \qquad (5.57)$$

Dividing the equation (5.56) by $m$, we have:

$$\frac{d^2x_1}{dt_1^2} + 2n_1\frac{dx_1}{dt_1} = r \qquad (5.58)$$

where $x_1$, $t_1$, and $n_1$ are the displacement, time, and damping factor for the first interval respectively, while

$$2n_1 = \frac{C_1}{m}$$

The Laplace domain of the equation (5.58) at the initial conditions (5.57) reads:

$$l^2x_1(l) + 2n_1lx_1(l) - r \qquad (5.59)$$

The solution of equation (5.59) in the Laplace domain is:

$$x_1(l) = \frac{r}{l(l + 2n_1)} \qquad (5.60)$$

The right side of equation (5.60) is not found in the Laplace Transform pairs of Table 3.1; however, a similar expression was analyzed that considered equation (3.62). Based on the method of decomposition, this equation was transformed into equation (3.63), which is represented in Table 3.1. So, following the example established with equation (3.63), we can transform equation (5.60) to the following form:

$$x_1(l) = \frac{r}{2n_1 l} - \frac{r}{2n_1[l-(-2n_1)]} \tag{5.61}$$

The inversion of the equation (5.61) into the time domain reads:

$$x_1 = \frac{r}{2n_1}t_1 + \frac{r(e^{-2n_1 t_1} - 1)}{4n_1^2} \tag{5.62}$$

Taking the first derivative from equation (5.62), we determine the velocity:

$$\frac{dx_1}{dt_1} = \frac{r}{2n_1}(1 - e^{-2n_1 t_1}) \tag{5.63}$$

At the end of the first interval, the velocity equals to $v_1$; thus, we can write:

$$v_1 = \frac{r}{2n_1}(1 - e^{-2n_1 T_1}) \tag{5.64}$$

where $T_1$ is the time duration of the first interval. According to equation (5.64), we determine:

$$e^{-2n_1 T_1} = \frac{r - 2n_1 T_1}{r} \tag{5.65}$$

and

$$T_1 = \frac{1}{2n_1} \ln \frac{r}{r - 2n_1 v_1} \tag{5.66}$$

Combining equations (5.62), (5.65), and (5.66), we determine the displacement $S_1$ on the first interval:

$$S_1 = \frac{r}{4n_1^2} ln \frac{r}{r - 2n_1 v_1} - \frac{v_1}{2n_1} \qquad (5.67)$$

The value of the displacement according to equation (5.67) and the value of velocity $v_1$ according to equation (5.61) represent the initial conditions of motion on the second interval.

### 5.2.2 Second Interval

The slope of line AB in Figure. 5.2 is proportional to the damping coefficient $C_2$ for the second interval. Considering triangle ABC, we may write:

$$C_2 = \frac{P_2 - P_1}{v_2 - v_1} \qquad (5.68)$$

However, the damping force on the second interval is not just the product of multiplying the damping coefficient $C_2$ by the velocity. According to Figure 5.2, the velocity $v_v$ causes a damping resistance force that equals the following sum of forces:

$$P_0 = P_1 + P_v \qquad (5.69)$$

where $P_v$ is the variable damping force on the second interval.

According to piece-wise linear approximation, the damping force on the second interval is structured as a sum of a constant force $P_1$ and a variable force $P_v$. Based on Figure 5.2, we may write the following expression:

$$P_v = C_2(v_v - v_1) \qquad (5.70)$$

Combining expressions (5.69) and (5.70), we have:

$$P_0 = P_1 + C_2 v_v - C_2 v_1 \qquad (5.71)$$

Actually, $P_1$ and $C_2 v_1$ are constant forces whereas $C_2 v_v$ is a variable force.

Thus, the differential equation of motion for the second interval should include the damping resisting force according to equation (5.71), whereas velocity $v_y$ should be replaced by the first derivative. Based on these considerations and referring to equation (1.6) we can compose the following differential equation of motion for the second interval:

$$m\frac{d^2x_2}{dt_2^2} + C_2\frac{dx_2}{dt_2} + P_1 - C_2v_1 = R \tag{5.72}$$

where $x_2$ and $t_2$ are respectively the displacement and time on the second interval.

The initial conditions of motion are:

$$\text{for} \quad t_2 = 0;\ x_2 = S_1; \quad \frac{dx_2}{dt_2} = v_1 \tag{5.73}$$

Dividing equation (5.72) by $m$, we have:

$$\frac{d^2x_2}{dt_2^2} + 2n_2\frac{dx_2}{dt_2} + p_1 - 2n_2v_1 = r \tag{5.74}$$

where

$$n_2 = \frac{C_2}{m} \tag{5.75}$$

The Laplace domain of equation (5.74) at the initial conditions (5.73) reads:

$$l^2x_2(l) - lv_1 - l^2S_1 + 2n_2lx_2(l) - 2n_2S_1l + p_1 - 2n_2v_1 = r \tag{5.76}$$

The solution of equation (5.76) in the Laplace domain is:

$$x_2(l) = \frac{r - p_1 + 2n_2v_1}{l(l + 2n_2)} + \frac{2n_2S_1 + v_1}{l + 2n_2} + \frac{lS_1}{l + 2n_2} \tag{5.77}$$

Applying to equation (5.77) similar actions as we did to equations (3.62), (3.63), and (5.61), we obtain:

$$x_2(l) = \frac{r - p_1 + 2n_2 v_1}{2n_2 l} - \frac{r - p_1 + 2n_2 v_1}{2n_2[l - (-2n_2)]} + \frac{2n_2 S_1 + v_1}{l - (-2n_2)}$$
$$+ \frac{lS_1}{l + 2n_2} \tag{5.78}$$

The inversion of equation (5.78) into the time domain reads:

$$x_2 = \frac{r - p_1 + 2n_2 v_1}{2n_2} t_2 + \frac{r - p_1 - 4n_2^2 S_1}{4n_2^2}\left(e^{-2n_2 t_2} - 1\right) + S_1 e^{-2n_2 t_2} \tag{5.79}$$

Taking the first derivative from equation (5.79), we can calculate the velocity:

$$\frac{dx_2}{dt_2} = \frac{r - p_1 + 2n_2 v_1}{2n_2} - \frac{r - p_1}{2n_2}e^{-2n_2 t_2} \tag{5.80}$$

Differentiating equation (5.80), we determine the acceleration:

$$\frac{d^2 x_2}{dt_2^2} = (r - p_1)e^{-2n_2 t_2} \tag{5.81}$$

At the end of the second interval, the vessel should reach its maximum velocity $v_v$; consequently, the acceleration at this moment should approach zero. Based on these considerations related to equation (5.81), we can write:

$$e^{-2n_2 T_2} \to 0 \tag{5.82}$$

where $T_2$ is the time duration of the motion on the second interval.

Combining equations (5.80) and expression (5.82), and considering that at the end of the second interval the velocity approaches $v_v$, we write:

$$v_v \to \frac{r - p_1 + 2n_2 v_1}{2n_2} \tag{5.83}$$

The only unknown parameter in expression (5.83) is the one associated with the active force:

$$r = \frac{R}{m}$$

Using the equality sign in expression (5.83) while inserting all related notations, we can determine the required active force $R$:

$$R = \frac{P_2\left(v_v - v_1\right) + P_1\left(v_2 - v_1\right)}{v_2 - v_1} \tag{5.84}$$

Multiplying the force according to equation (5.84) by the vessel's maximum velocity, we can determine the required power $N$ of the energy source that should be installed in the vessel:

$$N = Rv_v$$

This concludes the analysis of the problem associated with the vessel.

# DYNAMICS OF TWO-DEGREE-OF-FREEDOM SYSTEMS

The mechanical systems considered thus far include one movable mass, the motion of which is described by one differential equation. By implication, each of these systems possesses one degree of freedom. However, some mechanical systems include multiple masses connected to each other by specific links that allow their relative motion in relation to each other. These systems are called multiple-degree-of-freedom systems — the number of masses defines the number of degrees of freedom. The motion of these masses depends on each other. However, they may have different parameters of motion. Consequently, the motion of each of these individual masses is described by its differential equation of motion whereas these equations are combined into a group of differential equations characterizing the motion of the mechanical system. The number of differential equations of motion equals the number of masses in the mechanical system. We are limiting ourselves in this text to the analysis of two-degree-of-freedom systems that have the most practical importance for actual mechanical engineering problems.

There are just two types of links that make up the multiple-degree-of-freedom systems. These links are the hydraulic link (or dashpot) and the elastic link (or spring). Figure 6.1 illustrates a two-degree-of-freedom system comprising masses $m_1$ and $m_2$, which are linked by a dashpot $C$; the motion of these masses is described by the displacement functions $x_1$ and $x_2$ respectively. Figure 6.2 shows a two-degree-of-freedom system where the masses are connected by a spring $K$, whereas in Figure 6.3 the masses are connected by a dashpot $C$ and spring $K$ acting in parallel. In a two-degree-of-freedom system, the two masses can be linked by a combination of a group of dashpots or a group of springs, or by both of these groups. In these cases, both the damping coefficient and the stiffness coefficient should be determined the same way as we considered earlier for single mass systems.

In order to analyze the dynamics of a two-degree-of-freedom system, we must first compose a system of two simultaneous differential equations of motion.

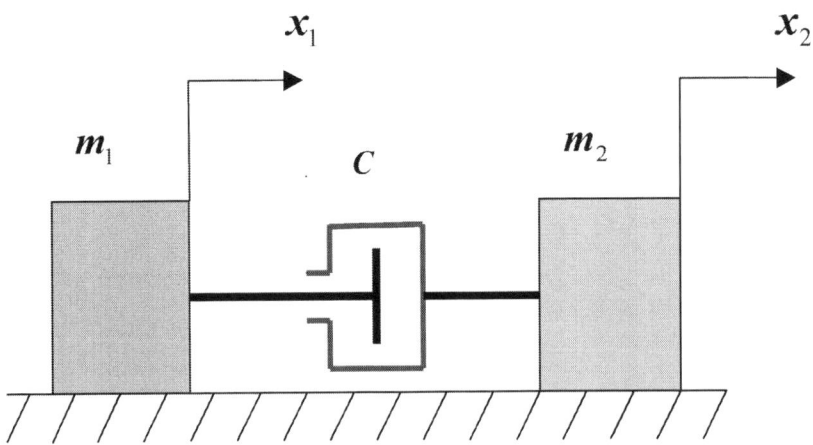

**Figure 6.1  A two-degree-of-freedom system with a dashpot.**

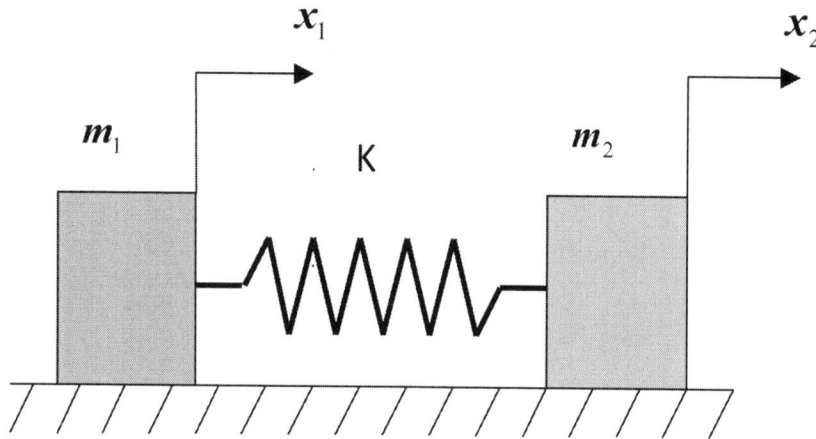

**Figure 6.2  A two-degree-of-freedom system with a spring.**

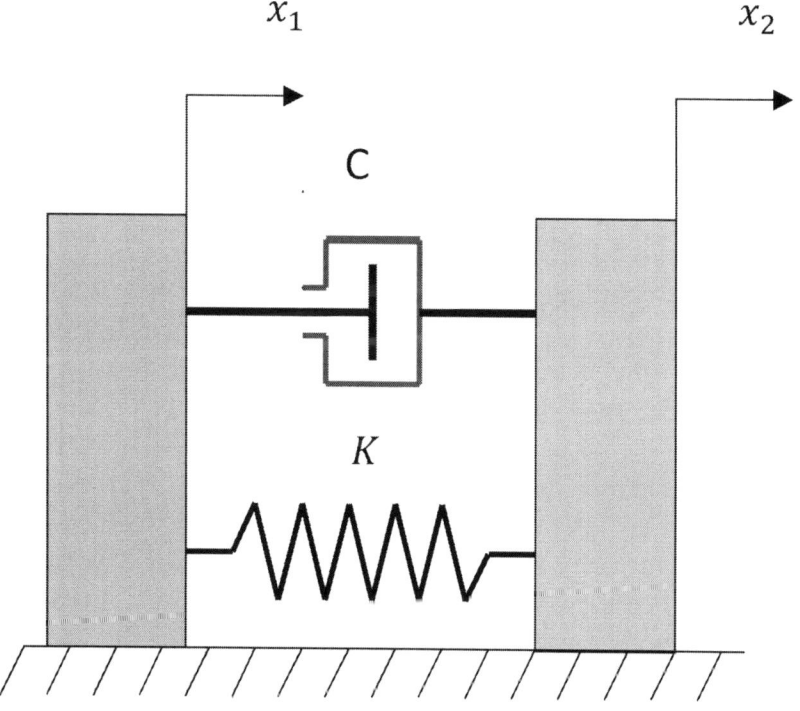

**Figure 6.3  A two-degree-of-freedom system with a dashpot and a spring.**

## 6.1  Differential Equations of Motion:  A Two-Degree-of-Freedom System

Three different two-degree-of-freedom systems are presented in Figures 6.1, 6.2, and 6.3. These three systems represent all possible combinations of linking two masses — there are no other types of two-degree-of-freedom systems. Each system should be characterized by an appropriate set of differential equations of motion. There are similarities in assembling these equations; however, it is more convenient to consider the systems separately.

### 6.1.1  A System with a Hydraulic Link (Dashpot)

Figure 6.1 illustrates a two-degree-of-freedom system where the two masses are linked by a dashpot. For simplicity, we assume that the masses move on a frictionless horizontal surface. Imagine that the first mass has a positive initial velocity due to an impact. The dashpot will immediately exert a resisting damping force to the first mass while the same force will start to push the second mass. However, the velocities of these two masses are not equal. (They can be equal just in cases when the link is a rigid body; in such cases, we obtain a system with one mass.)

In order to determine the actual resisting force applied to the first mass, we should first analyze the functioning of the dashpot. As seen in Figure 6.1, the first mass is rigidly connected to the piston whereas the second mass is connected to the cylinder. The piston, having the velocity of the first mass, pushes the fluid, which in its turn pushes the cylinder that moves together with the second mass as one body. Consequently, the net reaction of the dashpot being applied to both masses is proportional to the difference of the velocities between these two masses. The damping force that resists the motion of the first mass equals the product of multiplying the damping coefficient by the difference between the velocities of the first and second masses.

Based on these considerations, and assuming that no external forces are applied to the system, we can compose the following differential equation of motion for the first mass:

$$m_1 \frac{d^2 x_1}{dt^2} + C\left( \frac{dx_1}{dt} - \frac{dx_2}{dt} \right) = 0 \qquad (6.1)$$

The same force that resists the motion of the first mass causes the motion of the second mass; it can be considered as an external or active force applied to the second mass. These considerations let us compose the differential equation of motion for the second mass:

$$m_2 \frac{d^2 x_2}{dt^2} = C\left(\frac{dx_1}{dt} - \frac{dx_2}{dt}\right)$$

Restructuring this expression to the shape of equation (6.1), we have:

$$m_2 \frac{d^2 x_2}{dt^2} + C\left(\frac{dx_2}{dt} - \frac{dx_1}{dt}\right) = 0 \qquad (6.2)$$

Equations (6.1) and (6.2) represent the set of differential equations of motion for a two-degree-of freedom system when the first mass obtains a certain positive initial velocity.

The initial conditions for this case are:

$$\text{for} \quad t = 0, \quad x_1 = s_{01}; \quad \frac{dx_1}{dt} = v_{01}; \quad x_2 = s_{02}; \quad \frac{dx_2}{dt} = 0 \qquad (6.3)$$

Suppose the second mass receives a positive initial velocity. Both masses will start to move simultaneously; as before, their velocities are not equal. We can now see that the damping resisting force for the second mass equal the product of multiplying the damping coefficient by the difference of the velocities between the second and the first masses. Here the differential equation of motion for the second mass reads:

$$m_2 \frac{d^2 x_2}{dt^2} + C\left(\frac{dx_2}{dt} - \frac{dx_1}{dt}\right) = 0$$

This equation is the same as equation (6.2) for the second mass in the previous case.

The second term of this equation is actually the external active force applied to the first mass. Thus, the differential equation of motion of the first mass is:

$$m_1 \frac{d^2 x_1}{dt^2} = C\left(\frac{dx_2}{dt} - \frac{dx_1}{dt}\right) \qquad (6.4)$$

Equation (6.4) is identical to equation (6.1); consequently, the second case is described by the same set of differential equations developed for the first case. The difference between these two cases is in the initial conditions of motion. Here, the initial velocity of the first mass equals zero, whereas the second mass has a certain positive initial velocity.

The parameters of the initial conditions of motion do not influence the structures of these two differential equations of motion for the considered two-degree-of-freedom system. Equations (6.1) and (6.2) represent this system if no frictional or active forces are applied to the masses. However, when frictional resisting forces are present, they should be added to the left side of equations (6.1) and (6.2) respectively. If active forces are applied to the masses, they should be written in the right side of equations (6.1) and (6.2). Examples are considered in the following sections.

### 6.1.2 A System with an Elastic Link (Spring)

Figure 6.2 shows a two-degree-of-freedom system where the masses are connected by a spring having a stiffness coefficient $K$. The masses move on a horizontal frictionless surface. We assume that the first mass has a positive initial velocity. No active forces are applied to the masses. In such cases, both masses start to move simultaneously. Because the first mass moves faster than the second one, the spring becomes compressed, causing the second mass to move in the same direction. The spring exerts a resisting stiffness force to the first mass. This force is proportional to the spring deformation, which equals the difference between the displacements of the first and second masses.

Thus, the resisting stiffness force applied to the first mass equals the product of multiplying the stiffness coefficient by the difference of the displacements of the first and second masses. Based on these considerations, we can compose the differential equation of motion for the first mass:

$$m_1 \frac{d^2 x_1}{dt^2} + K(x_1 - x_2) = 0 \qquad (6.5)$$

The same stiffness force acts as an active force for the second mass; consequently, we may write:

$$m_2 \frac{d^2 x_2}{dt^2} = K(x_1 - x_2)$$

This equation can be rewritten in the following way:

$$m_2 \frac{d^2 x_2}{dt^2} + K(x_2 - x_1) = 0 \qquad\qquad \textbf{(6.6)}$$

Equations (6.5) and (6.6) represent the set of differential equations of motion for this example; expression (6.3) represents the initial conditions of motion.

Next, we assume that the second mass obtained a certain positive velocity. The second mass stretches the spring, pulling the first mass in the positive direction. The spring exerts a resisting force for the second mass while the same force acts as an active force to the first mass. This force equals the product of multiplying the stiffness coefficient by the difference between the displacements of the second and first masses. Based on these considerations, we can write the differential equations of motion for the second and first masses respectively:

$$m_2 \frac{d^2 x_2}{dt^2} + K(x_2 - x_1) = 0$$

$$m_1 \frac{d^2 x_1}{dt^2} = K(x_2 - x_1)$$

These two equations are identical to equations (6.5) and (6.6). Thus, the difference between these cases is reflected in the initial conditions of motion. The structure of the differential equations describing the motion of two masses linked by a spring is similar to the structure of the equations when the two masses are linked by a dashpot.

In a two-degree-of-freedom system where the masses are linked by a spring, the motion of the masses can be caused by this spring if it is deformed while the initial velocities of both masses equal zero. In this case, the initial displacement of one or both masses should be different from zero. It was emphasized earlier that the structure of the differential equations does not depend upon initial conditions of motion; therefore, equations (6.5) and (6.6) are also applicable to this case.

If one or both masses are subjected to active or additional resisting forces, they should be included in the differential equations of motion in a way that is similar to the one-degree-of-freedom systems.

### 6.1.3  A System with a Combination of a Hydraulic Link (Dashpot) and an Elastic Link (Spring)

A two-degree-of-freedom mechanical system with a combination of a dashpot and a spring acting in parallel is shown in Figure 6.3. We can apply the same considerations we apply for a dashpot or spring alone. Thus, we develop similar terms associated with the damping and stiffness resisting forces.

These considerations let us compose the following set of differential equations of motion for a two-degree-of freedom system in which the two masses are linked by a dashpot and a spring:

$$m_1 \frac{d^2 x_1}{dt^2} + C\left(\frac{dx_1}{dt} - \frac{dx_2}{dt}\right) + K\left(x_1 - x_2\right) = 0 \qquad (6.7)$$

and

$$m_2 \frac{d^2 x_2}{dt^2} + C\left(\frac{dx_2}{dt} - \frac{dx_1}{dt}\right) + K\left(x_2 - x_1\right) = 0 \qquad (6.8)$$

The initial conditions of motion for this case are arbitrary; in a most general case, these conditions can be expressed in the following way:

$$\text{for} \quad t = 0; \quad x_1 = s_{01}; \quad \frac{dx_1}{dt} = v_{01}; \quad x_2 = s_{02}; \quad \frac{dx_2}{dt} = v_{02}$$

When active forces and frictional resisting forces are present, they should be added respectively to equations (6.7) and (6.8).

## 6.2  Solutions of Differential Equations of Motion for Two-Degree-of-Freedom Systems

The methodology of the Laplace Transform is applicable to solving systems of linear differential equations. The examples below demonstrate the procedures required to solve sets of simultaneous differential equations of motion for two-degree-of-freedom mechanical systems.

### 6.2.1  Solutions for a System with a Hydraulic Link

Let us consider the motion of the system shown in Figure 6.1 in which the masses are moving on a frictionless surface in the horizontal direction while the first mass reaches a certain initial velocity. No active forces are applied to the masses. The system of differential equations of motion is already presented by the equations (6.1) and (6.2). The initial conditions of motion in this case are:

$$\text{for}\quad t=0;\ \ x_1=0;\ \ \frac{dx_1}{dt}=v_{01};\ \ x_2=0;\ \ \frac{dx_2}{dt}=0 \qquad (6.9)$$

Dividing equation (6.1) by $m_1$ and equation (6.2) by $m_2$, we have:

$$\frac{d^2 x_1}{dt^2}+2n_1\left(x_1-x_2\right)=0 \qquad (6.10)$$

$$\frac{d^2 x_2}{dt^2}+2n_2\left(x_2-x_1\right)=0 \qquad (6.11)$$

where

$$2n_1=\frac{C}{m_1} \qquad (6.12)$$

and

$$2n_2=\frac{C}{m_2} \qquad (6.13)$$

The respective Laplace domain of equations (6.10) and (6.11) at the initial conditions (6.9) are:

$$l^2 x_1(l) - l v_{01} + 2n_1 l x_1(l) - 2n_1 l x_2(l) = 0 \qquad (6.14)$$

$$l^2 x_2(l) + 2n_2 l x_2(l) - 2n_2 l x_1(l) = 0 \qquad (6.15)$$

Each of these equations (6.14) and (6.15) comprise the Laplace domain parameters $x_1(l)$ and $x_2(l)$. In order to solve the equation (6.14) for $x_1(l)$, we need to eliminate from it the term that comprises the parameter $x_2(l)$. We can do this based on the method of substitutions for a system of simultaneous equations. According to this method, we first solve equation (6.15) for $x_2(l)$, and we have:

$$x_2(l) = \frac{2n_2 x_1(l)}{l + 2n_2} \qquad (6.16)$$

Substituting $x_2(l)$ according to equation (6.16) into equation (6.14), we obtain:

$$l^2 x_1(l) - l v_{01} + 2n_1 l x_1(l) - \frac{4n_1 n_2 l x_1(l)}{l + 2n_2} = 0 \qquad (6.17)$$

Thus, equation (6.14) no longer includes the parameter $x_2(l)$, and consequently can be solved for $x_1(l)$:

$$x_1(l) = \frac{v_{01}(l + 2n_2)}{l[l + 2(n_1 + n_2)]} \qquad (6.18)$$

Denoting

$$n_0 = n_1 + n_2 \qquad (6.19)$$

we may write:

$$x_1(l) = \frac{v_{01}}{l + 2n_0} + \frac{2n_2 v_{01}}{l(l + 2n_0)} \qquad (6.20)$$

Performing with equation (6.20) the transformations similar to those we applied to equations (3.62) and (3.63), we have:

$$x_1(l) = \frac{v_{01}}{l - (-2n_0)} + \frac{v_{01}n_2}{n_0 l} - \frac{v_{01}n_2}{n_0[l - (-2n_0)]} \tag{6.21}$$

The inversion of equation (6.21) into the time domain reads:

$$x_1 = \frac{v_{01}n_2}{n_0}t - \frac{v_{01}n_1}{2n_0^2}(e^{-2n_0 t} - 1) \tag{6.22}$$

Continuing the method of substitutions, we solve equation (6.14) for $x_1(l)$ and we have:

$$x_1(l) = \frac{v_{01} + 2n_1 x_2(l)}{l + 2n_1} \tag{6.23}$$

Substituting equation (6.23) into equation (6.15), we obtain the solution for $x_2(l)$:

$$x_2(l) = \frac{v_{01}n_2}{n_0 l} - \frac{v_{01}n_2}{n_0[l - (2n_0)]} \tag{6.24}$$

Inverting equation (6.24) into the time domain, we write:

$$x_2 = \frac{v_{01}n_2}{n_0}t + \frac{v_{01}n_2}{2n_0^2}(e^{-2n_0 t} - 1) \tag{6.25}$$

Thus, equations (6.22) and (6.25) represent respectively the solutions for the system of differential equations of motion (6.1) and (6.2) at the initial conditions (6.9).

The respective velocities and accelerations of the two masses for this system of two-degree-of-freedom are:

$$\frac{dx_1}{dt} = \frac{v_{01}n_2}{n_0} + \frac{v_{01}n_1}{n_0}e^{-2n_0 t} \tag{6.26}$$

$$\frac{dx_2}{dt} = \frac{v_{01}n_2}{n_0} - \frac{v_{01}n_2}{n_0}e^{-2n_0 t} \tag{6.27}$$

$$\frac{d^2 x_1}{dt^2} = -2v_{01}n_1 e^{-2n_0 t} \tag{6.28}$$

$$\frac{d^2 x_2}{dt^2} = 2v_{01}n_2 e^{-2n_0 t} \tag{6.29}$$

Equations (6.22), (6.25), (6.26), (6.27), (6.28), and (6.29) let us analyze the dynamics of this two-degree-of-freedom system.

## 6.2.2 Solutions for a System with an Elastic Link

Figure 6.4 shows a two-degree-of-freedom system where the masses are linked by a spring having a stiffness coefficient $K$. The masses are moving in the horizontal direction on a frictional surface that exerts friction resisting forces $F_1$ and $F_2$. The first mass reaches a positive velocity $v_{01}$, which causes the system to move. We need to analyze the dynamics of this system. The solutions of the differential equations of motion of these masses will allow us to conduct the analysis.

Based on Figure 6.4 and equations (6.5) and (6.6), we can compose the following system of differential equations:

$$m_1 \frac{d^2 x_1}{dt^2} + K(x_1 - x_2) + F_1 = 0 \tag{6.30}$$

$$m_2 \frac{d^2 x_2}{dt^2} + K(x_2 - x_1) + F_2 = 0 \tag{6.31}$$

The initial conditions of motion are:

$$\text{for} \quad t = 0; \ x_1 = 0; \ \frac{dx_1}{dt} = v_{01}; \ x_2 = 0; \ \frac{dx_2}{dt} = 0 \tag{6.32}$$

Dividing equation (6.30) by $m_1$ and the equation (6.31) by $m_2$, we obtain the following system of equations:

$$\frac{d^2 x_1}{dt^2} + \omega_{01}^2 (x_1 - x_2) + f_1 = 0 \tag{6.33}$$

$$\frac{d^2 x_2}{dt^2} + \omega_{02}^2 (x_2 - x_1) + f_2 = 0 \tag{6.34}$$

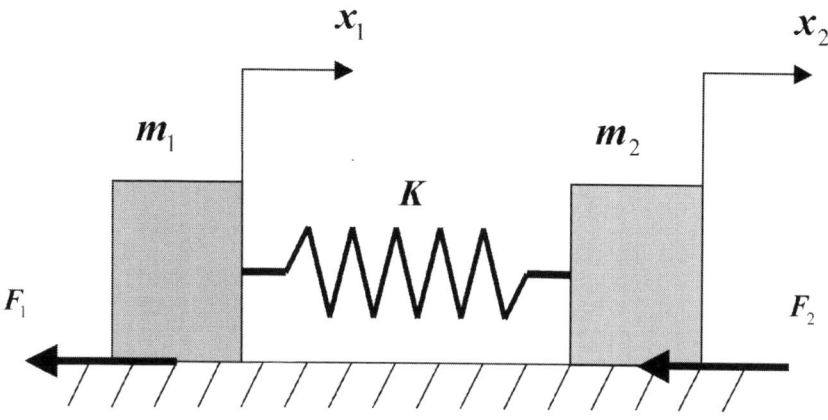

**Figure 6.4  A two-degree-of-freedom system is moving on a frictional surface; the masses are connected by a spring.**

where:

$$\omega_{01}^2 = \frac{K}{m_1} \tag{6.35}$$

$$\omega_{02}^2 = \frac{K}{m_2} \tag{6.36}$$

$$f_1 = \frac{F_1}{m_1} \tag{6.37}$$

$$f_2 = \frac{F_2}{m_1} \tag{6.38}$$

The Laplace domain of equations (6.33) and (6.34) at the initial conditions (6.32) and notations (6.35), (6.36), (6.37), and (6.38) respectively read:

$$l^2 x_1(l) - l v_{01} + \omega_{01}^2 x_1(l) - \omega_{01}^2 x_2(l) + f_1 = 0 \tag{6.39}$$

$$l^2 x_2(l) + \omega_{02}^2 x_2(l) - \omega_{02}^2 x_1(l) + f_2 = 0 \tag{6.40}$$

Solving equation (6.40) for $x_2(l)$, we have:

$$x_2(l) = \frac{\omega_{02}^2 x_1(l)}{l^2 + \omega_{02}^2} - \frac{f_2}{l^2 + \omega_{02}^2} \qquad (6.41)$$

Substituting equation (6.41) into equation (6.39), we can obtain the solution for $x_1(l)$:

$$x_1(l) = \frac{lv_{01}}{l^2 + \omega_{12}^2} + \frac{v_{01}\omega_{02}^2}{l(l^2 + \omega_{12}^2)} - \frac{f_1}{l^2 + \omega_{12}^2} - \frac{f_1\omega_{02}^2 - f_2\omega_{01}^2}{l^2(l^2 + \omega_{12}^2)} \qquad (6.42)$$

where

$$\omega_{12}^2 = \omega_{01}^2 + \omega_{02}^2 \qquad (6.43)$$

Using pairs 8, 19, 7, and 20 respectively from Table 3.1, we invert equation (6.42) into the time domain:

$$x_1 = \frac{v_{01}}{\omega_{12}}\sin\omega_{12}t + \frac{v_{01}\omega_{02}^2}{\omega_{12}^2}\left(t - \frac{1}{\omega_{12}}\sin\omega_{12}t\right)$$
$$- \frac{f_1}{\omega_{12}^2}(1 - \cos\omega_{12}t) - \frac{f_1\omega_{02}^2 + f_2\omega_{01}^2}{\omega_{12}^4}(\cos\omega_{12}t - 1) \quad (6.44)$$
$$- \frac{f_1\omega_{02}^2 + f_2\omega_{01}^2}{2\omega_{12}^2}t^2$$

Performing some appropriate actions with equation (6.44) we obtain:

$$x_1 = \frac{v_{01}\omega_{01}^2}{\omega_{12}^3}\sin\omega_{12}t + \frac{\omega_{01}^2(f_1 - f_2)}{\omega_{12}^4}\cos\omega_{12}t + \frac{v_{01}\omega_{02}^2}{\omega_{12}^2}t$$
$$- \frac{f_1\omega_{02}^2 + f_2\omega_{01}^2}{2\omega_{12}^2}t^2 - \frac{\omega_{01}^2(f_1 - f_2)}{\omega_{12}^4} \qquad (6.45)$$

In order to solve equation (6.40), we first need to determine $x_1(l)$ from equation (6.39). Solving this equation for $x_1(l)$, we have:

$$x_1(l) = \frac{\omega_{01}^2 x_2(l)}{l^2 + \omega_{01}^2} + \frac{lv_{01}}{l^2 + \omega_{01}^2} - \frac{f_1}{l^2 + \omega_{01}^2} \qquad (6.46)$$

Substituting equation (6.46) into equation (6.40), we can obtain the following equation for $x_2(l)$:

$$x_2(l) = \frac{v_{01}\omega_{02}^2}{l(l^2 + \omega_{12}^2)} - \frac{f_1\omega_{02}^2 + f_2\omega_{01}^2}{l^2(l^2 + \omega_{12}^2)} - \frac{f_2}{l^2 + \omega_{12}^2} \tag{6.47}$$

Using pairs 19, 20, and 7 respectively from Table 3.1, we invert equation (6.47) into the time domain. In turn, we have:

$$\begin{aligned}
x_2 &= \frac{v_{01}\omega_{02}^2}{\omega_{12}^2}t - \frac{v_{01}\omega_{02}^2}{\omega_{12}^3}\sin\omega_{12}t - \frac{\omega_{02}^2(f_1 - f_2)}{\omega_{12}^4}\cos\omega_{12}t \\
&\quad - \frac{f_1\omega_{02}^2 + f_2\omega_{01}^2}{2\omega_{12}^2}t^2 + \frac{\omega_{02}^2(f_1 - f_2)}{\omega_{12}^4}
\end{aligned} \tag{6.48}$$

Taking the first derivatives from equations (6.45) and (6.48), we determine the velocities of the two masses respectively:

$$\begin{aligned}
\frac{dx_1}{dt} &= \frac{v_{01}\omega_{01}^2}{\omega_{12}^2}\cos\omega_{12}t - \frac{\omega_{01}^2(f_1 - f_2)}{\omega_{12}^3}\sin\omega_{12}t \\
&\quad + \frac{v_{01}\omega_{02}^2}{\omega_{12}^2} - \frac{f_1\omega_{02}^2 + f_2\omega_{01}^2}{\omega_{12}^2}t
\end{aligned} \tag{6.49}$$

$$\begin{aligned}
\frac{dx_2}{dt} &= \frac{v_{01}\omega_{02}^2}{\omega_{12}^2}(1 - \cos\omega_{12}t) + \frac{\omega_{02}^2(f_1 - f_2)}{\omega_{12}^3}\sin\omega_{12}t \\
&\quad - \frac{f_1\omega_{02}^2 + f_2\omega_{01}^2}{\omega_{12}^2}t
\end{aligned} \tag{6.50}$$

Differentiating equations (6.49) and (6.50), we obtain equations for the accelerations of the two masses respectively:

$$\begin{aligned}
\frac{d^2x_1}{dt^2} &= -\frac{v_{01}\omega_{01}^2}{\omega_{12}}\sin\omega_{12}t - \frac{\omega_{01}^2(f_1 - f_2)}{\omega_{12}^2}\cos\omega_{12}t \\
&\quad - \frac{f_1\omega_{02}^2 + f_2\omega_{01}^2}{\omega_{12}^2}
\end{aligned} \tag{6.51}$$

$$\frac{d^2x_2}{dt^2} = \frac{v_{01}\omega_{01}^2}{\omega_{12}}\sin\omega_{12}t + \frac{\omega_{01}^2(f_1 - f_2)}{\omega_{12}^2}\cos\omega_{12}t$$
$$- \frac{f_1\omega_{02}^2 + f_2\omega_{01}^2}{\omega_{12}^2}$$

$$(6.52)$$

Thus, equations (6.45), (6.48), (6.49), (6.50), (6.51), and (6.52) let us analyze the considered two-degrees-of-freedom system.

In this example, the friction forces instantaneously change their directions when the velocities change their directions. Therefore, this system of differential equations of motion is valid just to the moment of time when one of these masses stops. To continue analyzing this system, we must determine the parameters for the moment when one of the masses stops; using these values as the initial conditions for the successive system of differential equations of motion.

### 6.2.3 Solutions for a System with a Combination of a Hydraulic and an Elastic Link

An automobile's regular shock-absorbing mechanism represents a two-degree-of-freedom system where the wheel assembly is connected to the body of the vehicle by both a dashpot and a spring. Figure 6.5 illustrates this system, $m_1$ and $W_1$ are the mass and weight of the body shown in position B, and $m_2$ and $W_2$ are the mass and the weight of the wheel assembly respectively. Axes $x_1$ and $x_2$ are directed upward; the origin of the axis $x_1$ is aligned with the position A of the first mass in case the spring is not deformed. The negative displacement $-S_1$ equals the deformation of the spring under the weight of the first mass; it represents the initial displacement of this mass.

The impact on the wheel in the vertical direction that is caused by the irregularities of the road results in the wheel assembly gaining a certain initial velocity that activates the shock absorbing system. In Figure 6.5, the wheel assembly reaches a positive initial velocity $v_2$. Based on this model and also referring to equations (6.7) and (6.8), we can compose the following system of differential equations of motion for these two masses:

$$m_1\frac{d^2x_1}{dt^2} + C\left(\frac{dx_1}{dt} - \frac{dx_2}{dt}\right) + K(x_1 - x_2) + W_1 = 0 \qquad (6.53)$$

$$m_2 \frac{d^2 x_2}{dt^2} + C\left(\frac{dx_2}{dt} - \frac{dx_1}{dt}\right) + K(x_2 - x_1) + W_2 = 0 \qquad \textbf{(6.54)}$$

The initial conditions are:

$$\text{for} \quad t = 0; \quad x_1 = -s_1; \quad x_2 = 0; \quad \frac{dx_1}{dt} = 0; \quad \frac{dx_2}{dt} = v_2 \qquad \textbf{(6.55)}$$

Dividing equation (6.53) by $m_1$ and equation (6.54) by $m_2$ we have:

$$\frac{d^2 x_1}{dt^2} + 2n_{01}\left(\frac{dx_1}{dt} - \frac{dx_2}{dt}\right) + \omega_{01}^2(x_1 - x_2) + g = 0 \qquad \textbf{(6.56)}$$

$$\frac{d^2 x_2}{dt^2} + 2n_{02}\left(\frac{dx_2}{dt} - \frac{dx_1}{dt}\right) + \omega_{02}^2(x_2 - x_1) + g = 0 \qquad \textbf{(6.57)}$$

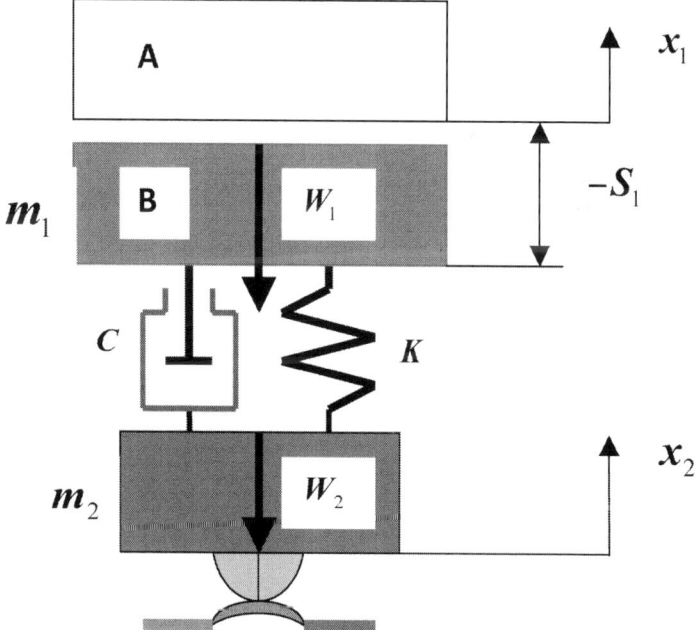

**Figure 6.5  A model of a two-degree-of-freedom shock-absorbing system.**

where:

$$2n_{01} = \frac{C}{m_1} \tag{6.58}$$

$$\omega_{01}^2 = \frac{K}{m_1} \tag{6.59}$$

$$2n_{02} = \frac{C}{m_2} \tag{6.60}$$

$$\omega_{02}^2 = \frac{K}{m_2} \tag{6.61}$$

and $g$ is the acceleration of gravity.

The Laplace domain of equations (6.56) and (6.67) at the initial conditions (6.55) respectively are:

$$l^2(l) + l^2 s_1 + 2n_{01}lx_1(l) + 2n_{01}ls_1 - 2n_{01}lx_2(l) \\ + \omega_{01}^2 x_1(l) - \omega_{01}^2 x_2(l) + g = 0 \tag{6.62}$$

$$l^2 x_2(l) - lv_2 + 2n_{02}lx_2(l) - 2n_{02}lx_1(l) + \omega_{02}^2 x_2(l) \\ - \omega_{02}^2 x_1(l) + g = 0 \tag{6.63}$$

In order to solve equation (6.62), we first need to determine $x_2(l)$ from equation (6.63). Solving equation (6.63) for $x_2(l)$, we have:

$$x_2(l) = \frac{x_1(l)(2n_{02}l + \omega_{02}^2)}{l^2 + 2n_{02}l + \omega_{02}^2} + \frac{lv_2 - g}{l^2 + 2n_{02}l + \omega_{02}^2} \tag{6.64}$$

Substituting $x_2(l)$ from equation (6.64) into equation (6.62), we can solve for $x_1(l)$:

$$x_1(l) = \frac{v_2\omega_{01}^2 - 2ng}{l\left[(l+n)^2 + \omega^2\right]} + \frac{2n_{01}v_2 - s_1\omega_{02}^2 - g}{(l+n)^2 + \omega^2} \\ - \frac{2n_{02}s_1 l}{(l+n)^2 + \omega^2} - \frac{s_1 l^2}{(l+n)^2 + \omega^2} - \frac{g\omega_{02}^2}{l^2[(l+n)^2 + \omega^2]} \tag{6.65}$$

where:

$$\omega^2 = \omega_{01}^2 + \omega_{02}^2 - (n_{01} + n_{02})^2$$

Consider the case where $\omega^2$ is positive. (We already considered when $\omega^2$ is negative or zero in Chapter 3.)

The last term of equation (6.65) is not represented in Table 3.1. Therefore, we will apply the method of decomposition in order to transform this term into components that are represented in this table. The decomposition reads:

$$\frac{1}{l^2[(l+n)^2 + \omega^2]} = \frac{A}{l} + \frac{B}{l^2} + \frac{C+Dl}{(l+n)^2 + \omega^2} \qquad (6.66)$$

Working further from (6.66), we have:

$$1 = Al^3 + 2nAl^2 + n^2Al + \omega^2 Al + Bl^2 + 2nBl$$
$$+ n^2B + \omega^2 B + Cl^2 + Dl^3 \qquad (6.67)$$

Based on equation (6.67), we can compose the following set of simultaneous equations that let us determine the unknown coefficients $A$, $B$, $C$, and $D$:

$$l^3(A+D) = 0$$
$$l^2(2nA + B + C) = 0$$
$$l(n^2A + \omega^2 A + 2nB) = 0$$
$$B(n^2 + \omega^2) = 1$$

Solving these four equations for the unknown coefficients, we obtain:

$$A = -\frac{2n}{(n^2 + \omega^2)^2} \qquad (6.68)$$

$$B = \frac{1}{n^2 + \omega^2} \qquad (6.69)$$

$$C = \frac{3n^2 - \omega^2}{(n^2 + \omega^2)^2} \qquad (6.70)$$

$$D = \frac{2n}{(n^2 + \omega^2)^2} \tag{6.71}$$

Substituting into equation (6.66) the values of the coefficients according to equations (6.68), (6.69), (6.70), and (6.71), we can re-write equation (6.65) in the following shape:

$$x_1(l) = \frac{v_2\omega_{01}^2 - 2ng}{l\left[(l+n)^2 + \omega^2\right]} + \frac{2n_{01}v_2 - s_1\omega_{02}^2 - g}{(l+n)^2 + \omega^2}$$

$$- \frac{2n_{02}s_1 l}{(l+n)^2 + \omega^2} - \frac{s_1 l^2}{(l+n)^2 + \omega^2} + \frac{2ng\omega_{02}^2}{l\left(n^2 + \omega^2\right)^2}$$

$$- \frac{g\omega_{02}^2}{l^2\left(n^2 + \omega^2\right)} - \frac{g\omega_{02}^2(3n^2 - \omega^2)}{\left(n^2 + \omega^2\right)^2\left[(l+n)^2 + \omega^2\right]}$$

$$- \frac{2ng\omega_{02}^2 l}{\left(n^2 + \omega^2\right)^2\left[(l+n)^2 + \omega^2\right]} \tag{6.72}$$

Using pairs 26, 13, 14, 15, 3, 4, 13, and 14 respectively from Table 3.1 — in the order of the terms presented in the right side of equation (6.72) — we invert this equation into the time domain:

$$x_1 = \frac{v_2\omega_{01}^2 - 2ng}{n^2 + \omega^2}[t - \frac{2n}{n^2 + \omega^2} + \frac{1}{\omega}e^{-nt}\sin(\omega t - \varphi)]$$

$$+ \frac{2n_{01}v_2 - s_1\omega_{02}^2 - g}{n^2 + \omega^2}[1 - e^{-nt}(\cos\omega t + \frac{n}{\omega}\sin\omega t)]$$

$$- \frac{2n_{02}s_1}{\omega}e^{-nt}\sin\omega t - s_1 e^{-nt}(\cos\omega t + \frac{n}{\omega}\sin\omega t)$$

$$+ \frac{2ng\omega_{02}^2}{(n^2 + \omega^2)^2}t - \frac{g\omega_{02}^2}{2(n^2 + \omega^2)}t^2 \tag{6.73}$$

$$- \frac{g\omega_{02}^2(3n^2 - \omega^2)}{\left(n^2 + \omega^2\right)^3}[1 - e^{-nt}(\cos\omega t + \frac{n}{\omega}\sin\omega t)]$$

$$- \frac{2ng\omega_{02}^2}{\omega(n^2 + \omega^2)}e^{-nt}\sin\omega t$$

where

$$\varphi = 2\tan^{-1}(-\frac{\omega}{n})$$

(6.74)

In order to calculate the solution for $x_2$, we need to perform similar actions as for $x_1$. Thus, we solve equation (6.62) for $x_1(l)$:

$$x_1(l) = \frac{x_2(l)\left(2n_{01}l + \omega_{01}^2\right)}{l^2 + 2n_{01}l + \omega_{01}^2} - \frac{l^2 s_1 + 2n_{01}ls_1 + g}{l^2 + 2n_{01}l + \omega_{01}^2}$$

(6.75)

Combining equations (6.63) and (6.75), we obtain the expression for $x_2(l)$:

$$x_2(l) = \frac{l(v_2 - 2n_{02}s_1)}{(l+n)^2 + \omega^2} + \frac{2n_{01}v_2 - g - 4n_{01}n_{02}s_1 - \omega_{02}^2 s_1}{(l+n)^2 + \omega^2}$$
$$+ \frac{v_2\omega_{01}^2 - 2ng - 2n_{01}\omega_{02}s_1}{l[(l+n)^2 + \omega^2]} - \frac{g\omega_{12}^2}{l^2[(l+n)^2 + \omega^2]}$$

(6.76)

The last term of equation (6.76) is similar to the last term of equation (6.65); the decomposition (6.66) is also applicable. Applying this decomposition to equation (6.76), we have:

$$x_2(l) = \frac{l(v_2 - 2n_{02}s_1)}{(l+n)^2 + \omega^2} + \frac{2n_{01}v_2 - g - 4n_{01}n_{02}s_1 - \omega_{02}^2 s_1}{(l+n)^2 + \omega^2}$$
$$+ \frac{v_2\omega_{01}^2 - 2ng - 2n_{01}\omega_{02}s_1}{l[(l+n)^2 + \omega^2]} + \frac{2ng\omega_{12}^2}{l(n^2 + \omega^2)^2}$$
$$- \frac{g\omega_{12}^2}{l^2(n^2 + \omega^2)} - \frac{g\omega_{12}^2(3n^2 - \omega^2)}{(n^2 + \omega^2)^2[(l+n)^2 + \omega^2]}$$
$$- \frac{2ngl\omega_{12}^2}{(n^2 + \omega^2)^2[(l+n)^2 + \omega^2]}$$

(6.77)

Using pairs 14, 13, 26, 3, 4, 13, and 14 from Table 3.1 — respectively in the order of the terms presented in the right side of equation (6.77) — we invert it into the time domain:

$$
x_2 = \frac{v_2 - 2n_{02}s_1}{\omega} e^{-nt} \sin \omega t + \frac{2n_{01}v_2 - g - 4n_{01}n_{02}s_1 - \omega_{02}^2 s_1}{\omega^2 + n^2}
$$

$$
\times [1 - e^{-nt}(\cos \omega t + \frac{n}{\omega} \sin \omega t)] + \frac{v_2 \omega_{01}^2 - 2ng - 2n_{01}\omega_{02}}{\omega^2 + n^2}
$$

$$
\times [t - \frac{2n}{\omega^2 + n^2} + \frac{1}{\omega} e^{-nt} \sin(\omega t - \varphi)] + \frac{2ng\omega_{12}^2}{(\omega^2 + n^2)^2} t \qquad (6.78)
$$

$$
- \frac{gw_{12}^2}{2(\omega^2 + n^2)} t^2 - \frac{g\omega_{12}^2(3n^2 - \omega^2)}{(\omega^2 + n^2)^2}
$$

$$
\times [1 - e^{-nt}(\cos \omega t + \frac{n}{\omega} \sin \omega t)] - \frac{2ng\omega_{12}^2}{\omega(\omega^2 + n^2)} \sin \omega t
$$

Differentiating equations (6.77) and (6.78) respectively, we can determine the velocities and accelerations of the masses. Actually, there is no need to present these parameters here; our goal is to demonstrate the step-by-step methodology of solving the system of differential equations of motion of a two-degree-of-freedom mechanical system where the masses are linked by a dashpot and a spring.

### 6.2.4 A System with a Hydraulic Link where the First Mass is Subjected to a Constant External Force

Figure 6.6 shows a model of a two-degree-of freedom system in which an active constant force $R$ is applied to the first mass. The masses are connected by a dashpot having a constant damping coefficient $C$; they are moving in the horizontal direction on a frictional surface that exerts friction forces $F_1$ and $F_2$ applied to the masses. The air resistance is ignored.

Based on Figure 6.6 and referring to equations (6.1) and (6.2), we can compose the following set of differential equations of motion for this system:

$$
m_1 \frac{d^2 x_1}{dt^2} + C \left( \frac{dx_1}{dt} - \frac{dx_2}{dt} \right) + F_1 = R \qquad (6.79)
$$

$$m_2 \frac{d^2 x_2}{dt^2} + C\left(\frac{dx_2}{dt} - \frac{dx_1}{dt}\right) + F_2 = 0 \tag{6.80}$$

The initial conditions are:

$$\text{for} \quad t = 0; \quad x_1 = 0; \quad x_2 = 0; \quad \frac{dx_1}{dt} = 0; \quad \frac{dx_2}{dt} = 0 \tag{6.81}$$

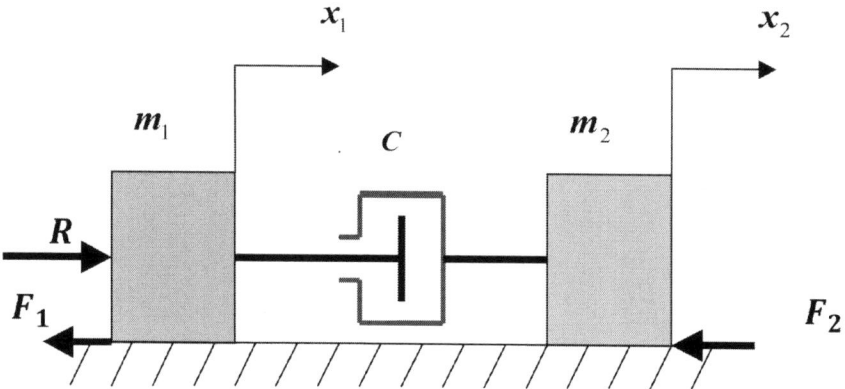

**Figure 6.6 A two-degree-of-freedom system with a dashpot, where the first mass is subjected to a constant external force.**

Dividing equation (6.79) by $m_1$ and equation (6.80) by $m_2$, we have:

$$\frac{d^2 x_1}{dt^2} + 2n_1\left(x_1 - x_2\right) + f_1 = r \tag{6.82}$$

$$\frac{d^2 x_2}{dt^2} + 2n_2\left(x_2 - x_1\right) + f_2 = 0 \tag{6.83}$$

where

$$f_1 = \frac{F_1}{m} \tag{6.84}$$

$$f_2 = \frac{F_2}{m} \tag{6.85}$$

$$r = \frac{R}{m} \qquad (6.86)$$

The Laplace domain of equations (6.82) and (6.83) at the initial conditions (6.81) are respectively:

$$l^2 x_1(l) + 2n_1 l x_1(l) - 2n_1 l x_2(l) + f_1 = r \qquad (6.87)$$

$$l^2 x_2(l) + 2n_2 l x_2(l) - 2n_2 l x_1(l) + f_2 = 0 \qquad (6.88)$$

In order to solve the equation (6.87), we need first to determine from equation (6.88) the Laplace domain displacement of the second mass $x_2(l)$:

$$x_2(l) = \frac{2n_2 l x_1(l) - f_2}{l(l + 2n_2)} \qquad (6.89)$$

Substituting equation (6.89) into equation (6.87), we eliminate the Laplace domain displacement $x_2(l)$ from it and determine the Laplace domain solution for the first mass $x_1(l)$:

$$x_1(l) = \frac{r - f_1}{l(l + 2n_0)} + \frac{2(rn_2 - f_2 n_1 - f_1 n_2)}{l^2(l + 2n_0)} \qquad (6.90)$$

Before inverting this expression into the time domain, we need to decompose the fractions in the right part of equation (6.90) — they are not found in Table 3.1. The first fraction is similar to the second fraction in equation (6.20); in turn, equation (6.21) shows how to resolve it into two fractions that are found in Table 3.1. Therefore, we will use this solution for the current case. The second fraction in equation (6.90) should undergo the decomposition process. Because the numerator of this fraction is a constant value, we can decompose the fraction in the following shape:

$$\frac{1}{l^2(l + 2n_0)} = \frac{A}{l} + \frac{B}{l^2} + \frac{C}{l + 2n_0} \qquad (6.91)$$

Working from equation (6.91), we develop the following equation:

$$1 = Al^2 + 2Aln_0 + Bl + 2Bn_0 + Cl^2$$

From this expression based on the powers of the parameter $l$, we compose three simultaneous equations for determining the coefficients $A$, $B$, and $C$. The values of these coefficients are:

$$A = -\frac{1}{4n_0^2}, \ B = \frac{1}{2n_0}, C = \frac{1}{4n_0^2}$$

Modifying equation (6.90) by recognizing its similarities with equation (6.20) and by using the decomposition results, we find:

$$
\begin{aligned}
x_1(l) = &\frac{n_1(r - f_1 + f_2)}{2n_0^2 l} + \frac{rn_2 - f_1 n_2 - f_2 n_1}{n_0 l^2} \\
&- \frac{n_1(r - f_1 + f_2)}{2n_0^2[l - (-2n_0)]}
\end{aligned}
\tag{6.92}
$$

All the fractions in the right side of equation (6.92) are represented in Table 3.1; thus, the inversion of this expression into the time domain reads:

$$
\begin{aligned}
x_1 = &\frac{n_1(r - f_1 + f_2)t}{2n_0^2} + \frac{(rn_2 - f_1 n_2 - f_2 n_1)t^2}{2n_0} \\
&+ \frac{n_1(r - f_1 + f_2)}{4n_0^3}(e^{-2n_0 t} - 1)
\end{aligned}
\tag{6.93}
$$

Continuing with the method of substitution, we determine the Laplace domain displacement for the first mass from equation (6.87):

$$x_1(l) = \frac{2n_1 l x_2(l) - f_1 + r}{l(l + 2n_1)} \tag{6.94}$$

Substituting equation (6.94) into equation (6.88), we eliminate the term $x_1(l)$ from it and determine the Laplace domain solution for the second mass $x_2(l)$:

$$x_2(l) = \frac{f_2}{l(l + 2n_0)} - \frac{2(rn_2 - f_2 n_1 - f_1 n_2)}{l^2(l + 2n_0)} \tag{6.95}$$

The decomposition that is applied to equation (6.90) can also be applied to equation (6.95) because they have similar structures. Thus, we transform equation (6.95) to the following shape:

$$x_2(l) = \frac{n_2(r - f_1 + f_2)}{2n_0^2 l} - \frac{rn_2 - f_1 n_2 - f_2 n_1}{n_0 l^2}$$
$$- \frac{n_2(r - f_1 + f_2)}{2n_0^2[l - (-2n_0)]} \tag{6.96}$$

The inversion of equation (6.96) into time domain reads:

$$x_2 = \frac{n_2(r - f_1 + f_2)t}{2n_0^2} - \frac{(rn_2 - f_1 n_2 - f_2 n_1)t^2}{2n_0}$$
$$+ \frac{n_2(r - f_1 + f_2)}{4n_0^3}\left(e^{-2n_0 t} - 1\right) \tag{6.97}$$

Equations (6.93) and (6.97) are respectively the solutions of the differential equations of motion. Taking the first derivatives from these equations, we can determine the velocities of the first and second masses respectively:

$$\frac{dx_1}{dt} = \frac{n_1(r - f_1 + f_2)}{2n_0^2} + \frac{(rn_2 - f_1 n_2 - f_2 n_1)t}{n_0}$$
$$- \frac{n_1(r - f_1 + f_2)}{2n_0^2}e^{-2n_0 t} \tag{6.98}$$

$$\frac{dx_2}{dt} = \frac{n_2(r - f_1 + f_2)}{2n_0^2} - \frac{(rn_2 - f_1 n_2 - f_2 n_1)t}{n_0}$$
$$- \frac{n_2(r - f_1 + f_2)}{2n_0^2}e^{-2n_0 t} \tag{6.99}$$

Differentiating equations (6.98) and (6.99), we determine the accelerations of these two masses:

$$\frac{d^2 x_1}{dt^2} = \frac{rn_2 - f_1 n_2 - f_2 n_1}{n_0} + \frac{n_1(r - f_1 + f_2)}{n_0}e^{-2n_0 t} \tag{6.100}$$

$$\frac{d^2 x_2}{dt^2} = -\frac{r n_2 - f_1 n_2 - f_2 n_1}{n_0} + \frac{n_2 \left( r - f_1 + f_2 \right)}{n_0} e^{-2 n_0 t} \qquad (6.101)$$

Equations (6.93), (6.97), (6.98), (6.99), (6.100), and (6.101) let us analyze the problem under consideration.

### 6.2.5 A Vibratory System Subjected to an External Sinusoidal Force

Figure 6.7 shows a two-degree-of-freedom vibratory system, where the masses are connected by a spring having a stiffness coefficient $K$. The masses are moving horizontally on a frictionless surface. The air resistance is negligible. The second mass is subjected to the action of sinusoidal force $A \sin \omega_0 t$, where $A$ is the amplitude of the force and $\omega_0$ is the frequency of the forced vibration.

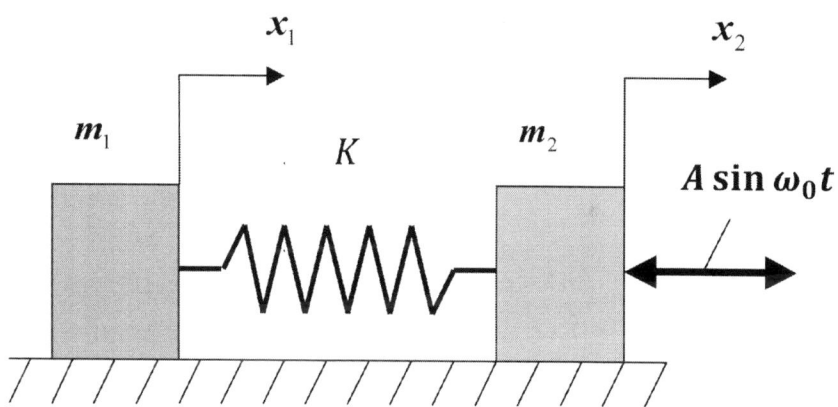

**Figure 6.7 A two-degree-of-freedom system with a sinusoidal external force.**

Based on Figure 6.7 and referring to equations (6.5) and (6.6), we can compose the following system of differential equations of motion for this case:

$$m_1 \frac{d^2 x_1}{dt^2} + K \left( x_1 - x_2 \right) = 0 \qquad (6.102)$$

$$m_2 \frac{d^2 x_2}{dt^2} + K(x_2 - x_1) = A \sin \omega_0 t \qquad (6.103)$$

The initial conditions of motion are found in expression (6.81).

Dividing equations (6.102) and (6.103) by $m_1$ and $m_2$ respectively, we have:

$$\frac{d^2 x_1}{dt^2} + \omega_1^2 (x_1 - x_2) = 0 \qquad (6.104)$$

$$\frac{d^2 x_2}{dt^2} + \omega_2^2 (x_2 - x_1) = a \sin \omega_0 t \qquad (6.105)$$

where

$$\omega_1^2 = \frac{K}{m_1} \qquad (6.106)$$

$$\omega_2^2 = \frac{K}{m_2} \qquad (6.107)$$

$$a = \frac{A}{m_2} \qquad (6.108)$$

The Laplace domain of equations (6.104) and (6.105) respectively are:

$$l^2 x_1 (l) + \omega_1^2 x_1 (l) - \omega_1^2 x_2 (l) = 0 \qquad (6.109)$$

$$l^2 x_2 (l) + \omega_2^2 x_2 (l) - \omega_2^2 x_1 (l) = \frac{a \omega_0 l}{l^2 + \omega_0^2} \qquad (6.110)$$

Applying the method of substitution, we determine from equation (6.110) the expression for the Laplace domain displacement of the second mass:

$$x_2 (l) = \frac{\omega_2^2 x_1 (l)}{l^2 + \omega_2^2} + \frac{a \omega_0 l}{(l^2 + \omega_0^2)(l^2 + \omega_2^2)} \qquad (6.111)$$

Substituting equation (6.111) into equation (6.109), we can determine the solution in Laplace domain for the displacement of the first mass:

$$x_1(l) = \frac{a\omega_0\omega_1^2}{l(l^2+\omega^2)(l^2+\omega_0^2)} \tag{6.112}$$

where

$$\omega^2 = \omega_1^2 + \omega_2^2 \tag{6.113}$$

The fraction in the right side of equation (6.112) is not found in Table 3.1. We decompose it into components that are represented in Table 3.1. Because the numerator has a constant value, we can decompose it in the following way:

$$\frac{1}{l(l^2+\omega^2)(l^2+\omega_0^2)} = \frac{A}{l} + \frac{Bl}{l^2+\omega^2} + \frac{Cl}{l^2+\omega_0^2} \tag{6.114}$$

Working from equation (6.114), we can write:

$$1 = Al^4 + Al^2\omega_0^2 + Al^2\omega^2 + A\omega^2\omega_0^2 + Bl^4$$
$$+ Bl^2\omega_0^2 + Cl^4 + Cl^2\omega^2$$

Using the method of undetermined coefficients for this expression, we compose a system of three simultaneous equations that allows to calculate the unknown coefficients:

$$1 = A\omega^2\omega_0^2$$
$$0 = Al^2\omega_0^2 + Al^2\omega^2 + Bl^2\omega_0^2 + Cl^2\omega^2,$$
$$0 = Al^4 + Bl^4 + Cl^4.$$

Solving these three equations simultaneously, we can determine the values of the coefficients:

$$A = \frac{1}{\omega^2\omega_0^2}$$

$$B = -\frac{1}{\omega^2(\omega_0^2 - \omega^2)}$$

$$C = \frac{1}{\omega_0^2(\omega_0^2 - \omega^2)}$$

Based on the decomposition, we modify the Laplace domain solution (6.112) into the following shape:

$$x_1(l) = \frac{a\omega_1^2}{\omega^2 \omega_0 l} - \frac{la\omega_0 \omega_1^2}{\omega^2(\omega_0^2 - \omega^2)(l^2 + \omega^2)}$$
$$+ \frac{la\omega_1^2}{\omega_0(\omega_0^2 - \omega^2)(l^2 + \omega_o^2)} \tag{6.115}$$

The inversion of equation (6.115) into the time domain reads:

$$x_1 = \frac{a\omega_1^2 t}{\omega^2 \omega_0} - \frac{a\omega_0 \omega_1^2}{\omega^3(\omega_0^2 - \omega^2)}\sin \omega t + \frac{a\omega_1^2}{\omega_0^2(\omega_0^2 - \omega^2)}\sin \omega_0 t \tag{6.116}$$

Equation (1.116) may undergo further transformation, but it is not necessary for this example.

In order to solve equation (6.110), we have to determine from equation (6.109) the Laplace domain expression for the displacement of the first mass $x_1(l)$. Performing the appropriate algebraic procedures with equation (6.109), we have:

$$x_1(l) = \frac{\omega_1^2 x_2(l)}{l^2 + \omega_1^2} \tag{6.117}$$

Substituting equaton (6.117) into equation (6.110), we can determine the Laplace domain solution for the second mass:

$$x_2(l) = \frac{a\omega_0 l}{(l^2 + \omega^2)(l^2 + \omega_0^2)} + \frac{a\omega_0 \omega_1^2}{l(l^2 + \omega^2)(l^2 + \omega_0^2)} \tag{6.118}$$

The first fraction in the right side of equation (6.118) is represented in Table 3.1 whereas the second fraction is identical with the right side of equation (6.112). Thus, we invert equation (6.118) into the time domain and obtain the solution of the differential equation of motion (6.105) for the second mass:

$$x_2 = \frac{a(\omega \sin \omega_0 t - \omega_0 \sin \omega t)}{\omega(\omega^2 - \omega_0^2)} + \frac{a\omega_1^2 t}{\omega^2 \omega_0}$$
$$- \frac{a\omega_0 \omega_1^2}{\omega^3(\omega_0^2 - \omega^2)}\sin \omega t + \frac{a\omega_1^2}{\omega_0^2(\omega_0^2 - \omega^2)}\sin \omega_0 t \tag{6.119}$$

Differentiating equations (6.116) and (6.119), we can determine the respective velocities of the first and the second masses:

$$\frac{dx_1}{dt} = \frac{a\omega_1^2}{\omega^2\omega_0} - \frac{a\omega_0\omega_1^2}{\omega^2(\omega_0^2 - \omega^2)}\cos\omega t + \frac{a\omega_1^2}{\omega_0(\omega_0^2 - \omega^2)}\cos\omega_0 t \quad \textbf{(6.120)}$$

$$\frac{dx_2}{dt} = \frac{a\omega_0(\cos\omega_0 t - \cos\omega t)}{\omega^2 - \omega_0^2} + \frac{a\omega_1^2}{\omega^2\omega_0}$$
$$- \frac{a\omega_0\omega_1^2}{\omega^2(\omega_0^2 - \omega^2)}\cos\omega t + \frac{a\omega_1^2}{\omega_0(\omega_0^2 - \omega^2)}\cos\omega_0 t \quad \textbf{(6.121)}$$

Taking the first derivatives from equations (6.120) and (6.121), we obtain equations of the accelerations for the first and second masses respectively:

$$\frac{d^2x_1}{dt^2} = \frac{a\omega_0\omega_1^2}{\omega(\omega_0^2 - \omega^2)}\sin\omega t - \frac{a\omega_1^2}{\omega_0^2 - \omega^2}\sin\omega_0 t \quad \textbf{(6.122)}$$

$$\frac{d^2x_2}{dt^2} = -\frac{a\omega_0(\omega_0\sin\omega_0 t - \omega\sin\omega t)}{\omega^2 - \omega_0^2}$$
$$+ \frac{a\omega_0\omega_1^2}{\omega(\omega_0^2 - \omega^2)}\sin\omega t - \frac{a\omega_1^2}{\omega_0^2 - \omega^2}\cos\omega_0 t \quad \textbf{(6.123)}$$

The equations for the displacements (6.116) and (6.119), for the velocities (6.120) and (6.121), and for the accelerations (6.122) and (6.123) allow us to completely analyze the two-degree-of-freedom mechanical system.

# INDEX

acceleration 2–3, 5, 10, 18, 71–76, 80, 82, 83–93, 95–100
active forces 5, 9–15, 29–31
air resistance 86
amplitude 11, 14
applied forces 29
approximation 115–136
argument 9
automobile 10, 83–100

backward stroke 108–110
barrel 95–100
braking system 71, 76–79, 84, 93–95

coefficient,
  damping 3
  stiffness 4
constant,
  force 8, 17, 29, 101–104, 158–163
  spring 4
conversion 34, 39, 73, 90, 94,105
cosine 38

damped motion 49–51, 56, 59–63
damping 63–65, 115
  coefficient 3
  first interval 131–133
  force 3, 8, 17, 19–22, 96
  non-linear 129–136
  resisting 80
  second interval 133–136
dashpots 3, 20–22, 138–142, 144, 145–148, 152–158
deceleration 71, 84
decomposition 157, 162
  fractions 40–45

deformation 24, 28, 125, 126, 128
  elastic media 4
  model 118
  plastic medium 5
degree-of-freedom systems 137
differential equation of motion 1–15, 70, 89, 102, 119, 122, 140–167
  inertia 46–49
  left side 2–9
  linear 4
  right side 5
  solution 1
  solving 46–67
displacement 2–3, 4, 5, 12, 13, 23, 27, 30–31, 33, 74–75, 103–104, 120, 127, 128
dry friction force 5, 8
dynamics 2, 5

elastic links 138-139, 142–144, 148–158
elastic media 4, 23–27
elasto-plastic models 23, 28, 117–129
electrical circuits 2
energy 6, 70
engineering systems 69–113
expansion 40
external forces 5, 17, 29, 60, 158–167
external loading factors 4

fluid medium 3, 19
fluid viscosity 15
force 17–31
  active 5, 29–31
  air resistance 86
  constant 8, 29, 101–104, 158–163

force (*continued*)
   damping  3, 7, 8, 17, 19–22
   dry friction  5, 8
   external  5, 17, 158–167
   friction  28–29, 77, 86
   gravity  5, 7, 13
   horizontal  86
   inertia  7, 8, 18–19
   pressure  95
   random  10–11
   reactive  17
   resisting  4, 7, 17–29
   sinusoidal  11, 13, 29
   stiffness  4, 7, 8, 23–27
   time  59–63
forced motion  63–65
forced vibrations  65–67
forward stroke  101–108
fractions
   decomposition  41–45
   proper rational  40–41
free vibrations  51–55
friction force  5, 17, 28–29, 77, 86,
     93, 96, 99. 115, 148
front-wheel drive  10
function  9

gravity  5, 7, 13, 27, 28, 78, 84
ground transportation  19, 83

harmonic function  11
horizontal axis  8–9, 13–14
horizontal forces  85
hydraulic links  3, 20, 138–142, 144,
     145–148, 152–163
hyperbolic cosine  38
hyperbolic sine  38, 58

impact  55–59
inertia  7, 8, 10, 17, 18–19
   damping  49–51

friction  48–49
   no resistance  46–48
   with resistance  48–51
initial velocity  104–108
inversion  34, 39, 73, 81, 94, 97, 103,
     106, 109, 112, 121, 123, 127, 132
     135, 150, 160, 162, 166

kinetic energy  6, 7, 46, 129

Laplace transforms  33–67, 70,
     81, 97, 102, 105, 109, 112,
     120, 123, 127, 131–132,
     134, 149,
     154, 160, 164
   conversion  34
   conversion  39
   inversion  34, 39
   inversion  39
   pairs  34–40, 73
law of motion  1
lifting, load  70–79
linear approximation  115–136
linear function  11–12
load, lifting  70–79
loading factors  9–15
   external  4

magnetism  27
mechanical system  1, 4, 5, 12, 14,
     21, 69–113
motion  5
   damping  49–51
   forced  63–65
   impact  55–59
   inertia  46–51
   parameters  70
   rectilinear  5–6
   vertical  13
multiple-degree-of-freedom
     systems  137

Newton, Second law  3
non-linear function  11–12
non-linearity  18, 116, 129–136

performance  70
piece-wise linear
    approximation  115–136
  first interval  119–122
  fourth interval  126–129
  second interval  122–124
  third interval  124–126
plastic medium  5
pneumatic operations  110–113
potential energy  6, 7
power needs  71
pressure forces  95
projectiles  95–100
proper rational fractions  40–41

random force  10–11
rational fractions  40–41
reactive forces  17
rear-wheel drive  10, 84–85
reciprocation cycle  100–110
rectilinear motion  5–6, 9, 12
resisting damping force  3, 7
resisting forces  17–29, 86, 119
  non-linear  129–136
resisting loading factors  9–15
rheological model  118
rotation  8–10

Second Law, Newton  3
ship  80–83
shock absorbers  21, 153
sine  38
sinusoidal force  11, 13, 29, 65–66,
    163–167
sliding link  100–110

soil penetrating machines  110–113
spring  51, 142–144, 148–158
spring constant  4
spring-loaded link  100–110
springs  4, 24–27, 138–139
steel cable  71
stiffness  7, 8, 63–65, 115, 117–119,
    122, 142–143
  coefficient  4
  force  4, 17, 23–27
stress-strain diagrams  24
systems, mechanical
    engineering  69–113

temperature  14–15
time  5, 11–14, 29, 59–63, 1
time domain  33–34, 39
transcendental equations  98
transform pairs  34–40
two-degree of freedom systems
    137–167

uniform motion  71, 83–84

velocity  2–3, 5, 12, 13, 28, 30, 70,
    73–74, 80, 82, 90–91, 99–100,
    103, 121, 127–128, 143
  initial  104–108
vertical motion  13
vibrations  51–55
  forced  65–67
vibratory systems  24–25, 28–29, 56,
    63, 65–66, 163–167
visco-elastic medium  116
visco-elasto-plastic models  23
viscosity  15

water vessel  80–83
wheel assembly  152